The Team Document

Ten years of leadership advancing the
National Occupational Research Agenda

U.S. Department of Health and Human Services
Public Health Services
Centers for Disease Control and Prevention
National Institute for Occupational Safety and Health

Washington, DC

April 2006

ORDERING INFORMATION

Copies of National Institute for Occupational Safety and Health (NIOSH) documents and information about occupational safety and health are available from

NIOSH-Publications Dissemination
4676 Columbia Parkway
Cincinnati, OH 45226-1998

Telephone:	1-800-35-NIOSH
	1-800-356-4674
Fax:	513-533-8573
E-mail:	pubstaft@cdc.gov
Webpage:	www.cdc.gov/niosh

This document is the public domain and may be freely copied or reprinted
Disclaimer: Mention of any company or product does not constitute endorsement by NIOSH.

DHHS (NIOSH) Publication No. 2006-121

CONTENTS

FOREWORD	V
INTRODUCTION	IX

DISEASE AND INJURY

Allergic and Irritant Dermatitis	3
Asthma and Chronic Obstructive Pulmonary Disease	9
Hearing Loss	15
Infectious Disease	21
Musculoskeletal Disorders (includes Low Back Disorders)	27
Reproductive Health Research (formerly Fertility and Pregnancy Abnormalities)	33
Traumatic Injuries	41

WORK ENVIRONMENT AND WORKFORCE

Emerging Technologies	49
Indoor Environment	55
Mixed Exposures	61
Organization of Work	67
Special Populations at Risk	73

RESEARCH TOOLS AND APPROACHES

Cancer Research Methods	81
Control Technology and Personal Protective Equipment	87
Exposure Assessment Methods	95
Health Services Research	101
Intervention Effectiveness Research	107
Risk Assessment Methods	113
Social and Economic Consequences of Workplace Illness and Injury	121
Surveillance Research Methods	127

APPENDIX

NORA Team and Related Publications	135
NORA Team Members	142

FOREWORD

Partnerships are vital to providing safe and healthy workplaces. Nowhere is this principle more realized than in the National Occupational Research Agenda, or NORA. Nearly ten years ago participants from diverse interests and perspectives joined NIOSH to establish a common research vision for the nation. This collaboration sparked a decade of leadership in occupational safety and health research.

Occupational injuries and illnesses affect us all. They result in losses of life, impairments in health, and diminished capacity for men and women in their prime. The burden these injuries and illnesses impose on families, communities, businesses, and the U.S. economy is enormous. Innovative research is critical for designing new tools and methods to reduce these burdens, and for anticipating new concerns in a changing workplace. No single agency or institution can face the challenges of mounting such research alone.

NORA offers a blueprint for developing effective partnerships. Through NORA diverse parties collaborated to produce innovative occupational safety and health research, and then worked to translate that research into effective workplace practices. By leveraging the talents and resources of many partners, NORA has stimulated important advancements in workplace safety and health.

As NORA marks its ten year anniversary NIOSH and our partners have the opportunity to reflect upon these historic public and private sector partnerships. I am pleased to present the Team Document, which describes both a decade of leadership in the 21 NORA priority areas and the teams' visions for the future of occupational safety and health research.

John Howard, MD
Director, National Institute for Occupational Safety and Health
Centers for Disease Control and Prevention

INTRODUCTION

Ten years ago, in the face of a rapidly expanding and increasingly diverse workplace, NIOSH asked the question:

What will the workplace of 2006 look like?

Significant progress had been made addressing many long-standing safety and health issues. Other hazards, however, remained problematic. And as the pace of technology advanced with unprecedented momentum, new challenges and opportunities were anticipated. Given this complex environment, what research would be needed to ensure safer, healthier workers in the 21st century?

The National Occupational Research Agenda, or NORA, sought to address these important questions. Unveiled in 1996, NORA became a map by which the occupational safety and health community could identify, generate, design, and fund priority research efforts. No previous occupational research agenda had captured such broad input and consensus. More than 500 individuals and organizations outside of NIOSH contributed to its development.

These contributions identified 21 priority areas for the 21st century. The priorities were not ranked, but did seek to encompass current and future needs. As illustrated below, the 21 priorities were grouped into three categories: Disease and Injury, Work Environment and Workforce, and Research Tools and Approaches.

Disease and Injury

Allergic and Irritant Dermatitis

Asthma and Chronic Obstructive Pulmonary Disease

Fertility and Pregnancy Abnormalities (later known as Reproductive Health Research)

Hearing Loss

Infectious Diseases

Low Back Disorders

Musculoskeletal Disorders

Traumatic Injuries

Work Environment and Workforce

Emerging Technologies

Indoor Environment

Mixed Exposures

Organization of Work

Special Populations at Risk

Research Tools and Approaches

Cancer Research Methods

Control Technology and Personal Protective Equipment

Exposure Assessment Methods

Health Services Research

Intervention Effectiveness Research

Risk Assessment Methods

Social and Economic Consequences of Workplace Illness and Injury

Surveillance Research Methods

The robust partnerships developed during this process fostered the creation of both the NORA Liaison Committee and the NORA research teams. The NORA Liaison Committee consisted of 22 diverse members who facilitated communication with stakeholders, encouraged participation on research teams, and identified opportunities to initiate research. Twenty priority research teams also were formed (two priority areas—musculoskeletal disorders of the upper extremities and low back disorders—were addressed by one team). These teams led the development and dissemination of new research under each NORA topic. Team membership included individuals from universities, professional organizations, major manufacturing industries, leaders in the insurance industry, health and safety professionals from organized labor, and representatives from several government agencies.

The diversity of participants contributed to the success of NORA, as the program effectively responded to its different constituencies. NORA team members fostered a unique forum to bridge gaps, encourage new research, and promote the adoption of effective workplace interventions. NORA enabled the Liaison Committee members to create relationships among organizations who had not previously collaborated. NORA supported occupational safety and health researchers in all 21 priority research areas, and offered a mechanism for funding organizations to focus their support in high-priority areas. Last, and somewhat unexpectedly, NORA became a partnership model for numerous local, state, and international occupational health organizations.

This document describes the successes of the teams and reflects on lessons learned during the first decade of NORA. Each NORA research team has described its efforts through a discussion of its priority area, progress made in the last ten years, and its perspective on important areas for future research. Advancing intervention effectiveness research, promoting new international standards for hearing-loss prevention, and describing new risks from hazardous drugs are just a few examples of impact that will be described in the following pages. Each chapter also includes a graph to illustrate trends in NORA funding by NIOSH during the past decade. A separate print compendium, titled *A Focus on Impacts: NORA Research 1996-2005*, and a more detailed electronic version titled *A Compendium of NORA Research Projects and Impacts, 1996-2005,* complement this volume and describe the accomplishments of individual NORA research projects.

Preventing injury and illness on the job, where working men and women spend an important part of their adult lives, is vital for sustaining workers and families, reducing health care costs, and maintaining a strong economy. NORA has made its mark for helping create safer, healthier workplaces. It is our hope that this document will foster reflection on our previous achievements and will help us move forward for another successful decade of occupational health safety research.

DISEASE AND INJURY

Allergic and Irritant Dermatitis

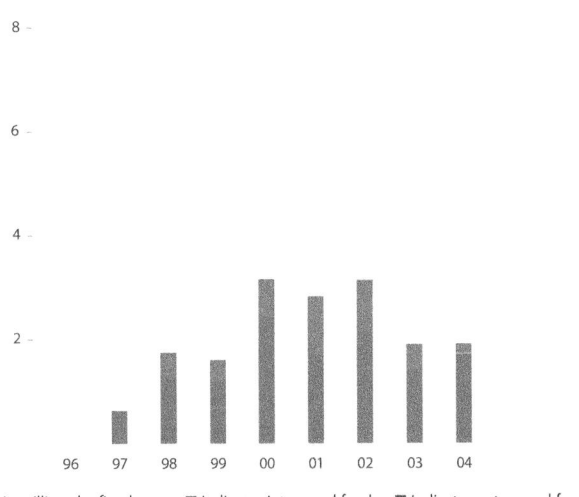

Funding in millions by fiscal year. ■ Indicates intramural funds ■ Indicates extramural funds

Few workplaces are free from exposures to the multitude of agents that cause allergic and irritant dermatitis. Prevention, therefore, is the key to reducing this common occupational disorder. Critical components of prevention include identifying the chemicals and proteins that cause allergic or irritant dermatitis, educating workers and their employers about the hazards of those chemicals, training physicians about occupational exposures, and promoting the benefits of good skin hygiene.

When NORA began in 1996, researchers had identified the most allergenic and irritant chemicals and proteins by either clinical experience or animal testing. Then, as now, most hazard identification data on skin sensitizers and irritants were the property of companies manufacturing consumer products, and little interest existed in researching occupational dermatitis. Laboratory researchers had few opportunities to exchange ideas with their clinical counterparts. One exception was the Dermal Exposure Network, organized and funded by the European Union to study occupational skin exposure and to develop better exposure assessment and risk assessment methods. This project was designed to improve the identification and prevention of systemic toxicity resulting from dermal chemical exposures. Elements of the European Dermal Exposure Network would later influence the research agenda of the NORA Allergic and Irritant Dermatitis (AID) team.

A DECADE OF PROGRESS

The team began with a multifaceted approach. First, they identified research gaps and priorities, assigning them to one of three groups: Basic Biomedical Science; Clinical/Epidemiology/Surveillance; and Exposure and Risk Assessment/Prevention. Although modified slightly, the list was incorporated into the FY 1998 Request for Application (RFA) for regular and pilot research grants. This pioneering collaboration joined NIOSH and the National Institutes of Health (NIH) to fund research projects in response to the NORA priorities. Five research projects were funded in 1998. In subsequent years, additional AID extramural projects were funded, including a surveillance project sponsored by the NORA Surveillance Research Methods priority area.

The team played a pivotal role in creating forums to share research results across disciplines and to plan improved prevention efforts. Team members were instrumental in forming the Experimental Contact Dermatitis Research Group (ECDRG). This group sponsors biannual conferences to discuss the basic and applied science of contact dermatitis. Scientists from academia, government, and industry exchange information with laboratory scientists and clinicians.

The team worked closely with several related professional organizations such as the American Contact Dermatitis Society and the American Industrial Hygiene Conference and Exposition to highlight recent research and clinical study results related to occupational skin diseases. Team members offered lectures and courses in a number of existing forums, such as the "UAW–GM Health and Safety Training Conference" (2001) and the "First International Symposium on Diagnosis, Treatment, and Prevention of Dermatological Problems Among Health Care Workers" (2001).

In response to the priority research areas identified by the team, NIOSH created the NORA Dermal Exposure Research Program. In 2000, 10 projects in this Program were initiated. The Program's accomplishments include research on chemical penetration into the skin for better risk assessment; research to develop improved quantitative structure-activity relationships for screening chemicals as possible skin sensitizers; development of information for occupational hygienists to guide investigations of occupational dermal issues; development of improved NIOSH policy on skin notations to identify potential skin irritants/corrosives, sensitizers, and systemic toxics by the skin route; and organization of two international conferences. The "Occupational and Environmental Exposures of Skin to Chemicals" conferences brought together occupational and environmental health professionals, dermatologists, laboratory scientists, policy makers and others to focus on improved prevention of local and systemic injury and disease caused by exposing skin to chemicals. In addition, the NIOSH Website features a Skin Topic Page to highlight documents and links to other useful resources, including the updated educational program entitled "Occupational Dermatoses—A Program for Physicians."

Finally, NORA-related research efforts have made positive contributions in understanding decontamination, sampling, worksite effectiveness of protective clothing, immunological responses leading to occupational skin diseases, and detection of surface contamination. The Occupational Safety and Health Administration (OSHA) has added important information related to dermal exposures to its Website, and the U.S. Environmental Protection Agency (EPA) has refined its guidance for assessing the hazard of skin exposures to chemicals in soil and water at hazardous waste sites.

FUTURE DIRECTIONS

During the next decade, more effort will be needed to compile information on successful interventions aimed at preventing hazardous exposures to the skin and to demonstrate the efficiencies resulting from such effective efforts. Increased awareness of occupational skin exposures and effects is still needed for dermatologists, occupational physicians, and occupational hygienists. Although there are numerous opportunities for these health professionals to take courses or read books about the identification and prevention of hazardous occupational skin exposures, recent efforts have not demonstrated an increased awareness of the potential problems. As a result, many opportunities to prevent harm are not recognized.

The team's research agenda highlights many areas in which research is needed:

Basic Biomedical Sciences

> Establish predictive tests (e.g., in vitro, QSAR, animal, and clinical) for identifying corrosive and irritant chemicals and allergens in the workplace.
>
> Develop a chemical and biological database (gold standards) for validation of a new methodology for irritant contact dermatitis (ICD) and allergic contact dermatitis (ACD).
>
> Develop approaches for assessing the irritant potential of chemical mixtures.
>
> Investigate susceptibility factors (e.g., genetic, age, metabolism, and environmental) influencing the development of ICD and ACD.
>
> Increase knowledge of basic pathophysiology of ICD and ACD.
>
> Encourage the exchange of information (e.g., scientific meetings) focusing on AID in the workplace.

Clinical/Epidemiology/Surveillance

Investigate the relationship of AID and respiratory disorders (epidemiology and pathogenesis).

Develop and validate assays for contact urticaria (immune and non-immune).

Develop and validate in vitro assays for ACD (cell-mediated immunity).

Develop and validate methods to identify precursors or predictors of clinical disease.

Increase knowledge of risk factors, prevalence, incidence, economic impact, natural history, prognosis, treatment, and prevention of hand dermatitis and dermatitis on other sites.

Improve, expand, and validate the methods to acquire medical and public health surveillance data with a focus on high-risk occupations and tasks. Support and expand current systems and programs.

Improve methods to study outbreaks by developing and validating protocols (to be conducted prior to more detailed NIOSH or other investigations).

Exposure and Risk Assessment/Prevention

Identify high-risk occupational groups and work tasks with potential for dermal exposure absorption.

Develop health-based criteria for acceptable occupational skin exposure to limit effects of irritancy, sensitization, or systemic toxicity.

Develop improved methods for measuring skin exposure.

Improve methods for assessing skin barrier properties.

> Improve risk assessment methods to evaluate AID and systemic toxicity.
>
> Improve existing and develop new preventive strategies and tools to reduce AID and systemic toxicity.
>
> Develop and evaluate approaches to training workers to prevent AID and systemic toxicity.

On the basis of funding and the sheer numbers of involved professionals, Europe is anticipated to remain the leader in skin disorders of occupational origin. Americans, however, are expected to make major contributions, with the hope that they will apply the integrated international knowledge learned to the pursuit of solving occupational skin exposure and disease problems. The continued emphasis on multidisciplinary and international forums to share information is essential to effective use of the resources allocated to identify and solve dermal exposure problems in the U.S. and internationally.

Asthma and
Chronic Obstructive Pulmonary Disease

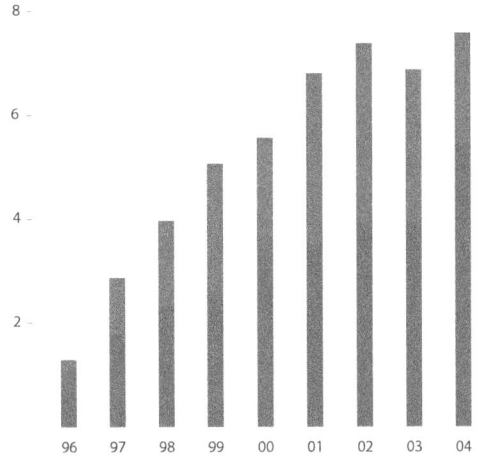

Funding in millions by fiscal year. ■ Indicates intramural funds ■ Indicates extramural funds

When NORA began in 1996, occupational asthma was the most frequent respiratory diagnosis among patients in occupational medicine clinics. Asthma and Chronic Obstructive Pulmonary Disease (COPD) caused nearly 92,000 deaths in the U.S. in 1992, making these airway diseases the fourth leading cause of mortality. Over a decade later, COPD continues to be a common outcome among workers exposed to dusts.

A DECADE OF PROGRESS

When the Asthma/COPD NORA Team formed in 1996, multiple lists of research goals existed for these conditions. Rather than contribute to this already robust literature, the team chose instead to encourage the implementation of the following goals:

> Determine the occupational contribution to the burden of asthma and COPD.
>
> Determine the economic impact of work-related asthma and COPD.
>
> Develop methods for the identification of work-related asthma.
>
> Determine the contribution of work to the exacerbation of asthma.
>
> Investigate the contribution of indoor air in non-industrial worksites to the onset and exacerbation of asthma.

Progress on these goals has been steady. First, team-supported research suggests that 15% to 19% of asthma and COPD can be attributed to workplace exposures. The team provided support to the American Thoracic Society (ATS) to study this issue. The ATS later published a statement concluding that 15% of asthma and COPD can be attributed to work. Work conducted by NIOSH, using data from the Third National Health and Nutrition Examination Survey (NHANES III), found that 19% of COPD was occupationally related. This estimate was as high as 31% among people who never smoked.

In addition to these serious health consequences, work-related asthma and COPD impose a substantial financial burden to the U.S. A team-supported analysis concluded that work-related asthma and COPD cost the nation $6.6 billion annually.

Recognizing these conditions to improve prevention and treatment remained an important priority. The team contributed to the development of methods that clinicians and researchers can use to identify work-related asthma. In addition, the team contributed funding and personnel to revise the ATS Respiratory Questionnaire. The new document will include more questions about asthma and work than the original survey.

Work within NIOSH also has been steady, with the launch of the Research on Occupational Asthma Reduction (ROAR) program in 2000. The individual projects within the ROAR program were:

> "Research for Occupational Asthma Reduction/Coordination";
>
> "Workplace Exacerbation of Asthma";
>
> "Work-Related Asthma in School and Office Buildings";
>
> "Medical Monitoring for Workers Using Isocyanates"; and
>
> "Lab Core for the Research on Occupational Asthma Reduction Program."

Several ROAR projects have demonstrated impressive impacts. Members of the "Workplace Exacerbation of Asthma" project collaborated with the manufacturer of a portable spirometer to develop software that facilitates testing for a work-related pattern in serial peak expiratory flow measurements. This software is now available to the manufacturer's customers at no additional cost. The project "Medical Monitoring for Workers Using Isocyanates" has continued to forge collaborative relationships with companies to investigate the effectiveness of medical monitoring among workers at risk for asthma. The ROAR project "Work-Related Asthma in School and Office Buildings" has resulted in the development of a visual assessment tool for documenting water damage and mold in buildings, increasing exposure assessment methods for non-industrial indoor working environments. Ratings assigned by the tool have been shown to have a positive association with questionnaire data for occupant health effects. Measured exposures from field studies in offices and schools are providing strong evidence for using dust as a historical marker of building contamination. Several components of dust, such as total and culturable fungi, endotoxin, and cat or dog allergens, are showing positive associations with occupant health effects. Five Health Hazard Evaluation Reports have been completed as part of this project.

Finally, baseline data from the ROAR project "Workplace Exacerbation of Asthma" and population-based studies in Colorado and Maine demonstrated that work does contribute to the exacerbation of symptoms in approximately 25% of adults with asthma. A 1998 commentary published in the *American Journal of Industrial Medicine* proposed that work-exacerbated asthma was potentially as serious as work-initiated asthma, since in both instances continued exposure could result in fixed airflow obstruction.

Many partnerships have resulted from the efforts of the Asthma/ COPD NORA Team, including partnerships with the National Heart, Lung, and Blood Institute (NHLBI) and the National Center for Environmental Health (NCEH). The NCEH partnership includes a joint study to address how home and work environments contribute to asthma onset, and an agreement to investigate indoor air quality and asthma in schools.

FUTURE DIRECTIONS
The team recommends the following activities for future NORA research:

- Evaluate the use of exhaled condensates for identifying work-related asthma.
- Investigate the occurrence and mechanisms of irritant and non-allergenic work-related asthma agents.
- Improve the methods for identifying people in early stages of rapid decline in pulmonary function.
- Conduct additional research on industries and occupations recently identified to pose a risk for COPD.
- Conduct intervention research for the prevention of work-related asthma and COPD.
- Document how genetic factors modify the effect of occupational exposures associated with asthma and COPD.
- Support the cognitive review and field testing of the Revised ATS Respiratory Questionnaire.

Hearing Loss

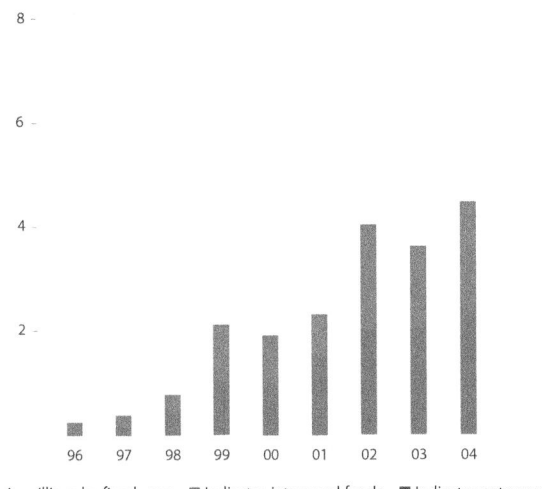

Funding in millions by fiscal year. ■ Indicates intramural funds ■ Indicates extramural funds

Occupational hearing loss is a pervasive problem, affecting people who work in manufacturing, construction, transportation, agriculture, services, and the military. Approximately 30 million American workers are exposed to potentially hazardous noise levels of 85 decibels or higher. At OSHA's present noise-exposure limit of 90 decibels, one in four people will develop a permanent hearing loss as a result of workplace exposures. In addition, more than 9 million American workers are exposed to solvents, metals, asphyxiates, and pesticides, which alone or in combination can damage their hearing.

Hearing loss denies individuals sensory experiences that contribute to the quality of life and impedes their ability to be gainfully employed. The gradual progression of hearing loss due to noise may be less dramatic than a work-related injury, but it is a significant and permanent handicap. These tragedies are also 100% preventable.

A DECADE OF PROGRESS

The NORA Hearing Loss Team has worked to identify and prioritize research needs to fill existing research gaps. The team pursued collaborations with academia, various industries, scientific organizations, and organized labor to increase awareness about noise and hearing loss prevention. The team sponsored two special sessions at the American Industrial Hygiene Conference and Exposition and one at the National Hearing Conservation Association (NHCA) in 2003. Separate from the special sessions, the Hearing Loss Team also developed a series of four Best Practices Workshops, each focused on an industry sector or a particular problem related to the prevention of occupational hearing loss.

The first workshop, held in 1999, focused on the Best Practices in Hearing Loss Prevention in the Manufacturing Sector. It was co-sponsored by NIOSH, NHCA, and Wayne State University. A second workshop a year later focused on the construction sector. Hosted by NIOSH, OSHA, and the Laborers' International Health and Safety Fund of North America (LHSFNA), the workshop prompted a renewed effort toward noise control research in construction at NIOSH, including development of a NIOSH Website describing the noise levels of common construction tools and information on available control technologies.

In 2002, the team partnered with NHCA to host a Best Practices Workshop on Combined Effects of Chemicals and Noise on Hearing. NIOSH published a report documenting the consensus reached during the workshop, and an overview of the current state of knowledge on the effects of industrial chemicals on the auditory system was published in 2003. This workshop had several important impacts. NIOSH funded a project, "Preventing Hearing Loss from Chemical and Noise Exposures." The American College of Occupational and Environmental Medicine referenced the conference report in its evidence-based statement on noise-induced hearing loss. The U.S. Army

Center for Health Promotion and Preventive Medicine used the information in its fact sheet titled Occupational Ototoxins (Ear Poisons) and Hearing Loss. In February 2003, the European Parliament published the Directive (2003/10/EC) on minimum health and safety requirements regarding the exposure of workers to the risks arising from noise. In Article 4 of Section II, Obligations of Employers, the Directive states that when carrying out risk assessments, employers should "…give particular attention to any effects on workers' health and safety resulting from interactions between noise and work-related ototoxic substances…" Member European countries have until 2006 to start the implementation of this new directive.

Finally, the fourth workshop of this series discussed Impact-Type Noise. Again co-sponsored by NHCA in 2003, it produced a document detailing the workshop consensus and an overview of the knowledge related to the effects of impact noise on the auditory system. This paper was submitted for publication to the *Noise Control Engineering Journal*, and NIOSH plans to publish the proceedings of this workshop. After the workshop, in October 2004, NIOSH started funding the project "New Methods for the Evaluation of Impact Noise Exposure."

Finally, NIOSH and the National Institute on Deafness and Other Communication Disorders co-sponsored an RFA relevant to the area of hearing loss prevention with particular focus on:

> The biological effects and biological responses to noise and other substances that damage hearing. Areas of interest included the relationship of first or second hand smoke to noise- induced hearing loss; the relationship of noise induced hearing loss to secondary risk factors such as hypertension, head trauma, diabetes, blood lipids, and clinical therapeutic drugs; the role of genetics in determining individual susceptibility to noise induced hearing loss; and the role of free radicals and antioxidants in noise induced hearing loss.

New noise control and personal protective equipment technologies. These technologies include active noise reduction control at the source; incorporation of electroacoustic systems in hearing protectors; and the development of methods for determining the actual noise reduction workers receive from hearing protectors.

Issues for hearing impaired workers. Of interest were the susceptibility of people with pre existing noise or age related hearing loss to additional noise induced hearing loss; personal protective equipment, such as hearing aids, for noise reduction; rehabilitation for hearing impaired workers who must continue to work in noise; and defining of audiometric and performance specifications for hearing critical jobs.

Effects of noise on speech communication. Issues include accuracy of communication in terms of expected outcomes or the need to repeat messages; and the safety of workers who must work and communicate in noise.

Health communications research. This area of interest includes methods for delivering training and motivation to noise exposed workers.

Surveillance and intervention research. Topics include evaluation methods to study the effectiveness of existing compliance driven hearing conservation programs, as well as development of longitudinal audiometric databases for persons not exposed to noise or other hazardous agents.

FUTURE DIRECTIONS

The team recommends developing model curricula for professional training programs in both audiology and industrial hygiene. Neither of these professions has specific training requirements for exposure assessment, application of controls, or how to provide an effective hearing loss prevention program. The curricula proposal should be presented to the AIHA, the American Speech-Language-Hearing Association, and the American Academy of Audiology. These associations accredit training programs and certify their respective professional members, yet none provides any guidance on course content for training in hearing loss prevention.

The team's sponsorship of lectures, sessions in conferences, and Best Practices Workshops has proven highly successful. This activity should continue, with emphasis on global outreach, currency of the information, and co-sponsorship with scientific and professional organizations. Future best practices and state-of-the-art conferences addressing small business, as well as new strategies for protecting miners from hearing loss, are recommended.

Infectious Diseases

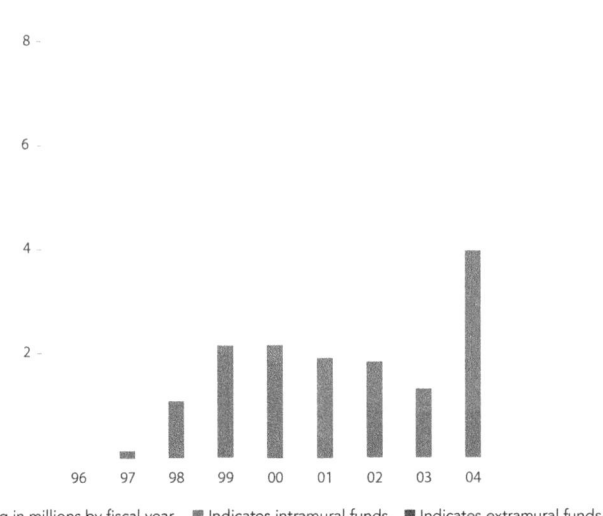

Funding in millions by fiscal year.   ■ Indicates intramural funds   ■ Indicates extramural funds

The last ten years have seen remarkable and unanticipated developments related to infectious diseases. The emergence of severe acute respiratory syndrome (SARS), avian influenza, monkeypox, and West Nile Virus demonstrated how quickly infectious diseases can become world-wide problems. The anthrax attacks of 2001 demonstrated the safety and economic threats posed by biological warfare agents. Postal workers and other unexpected groups were suddenly appreciated to be at risk for occupational infectious diseases.

When NORA began, health care workers were identified as a key occupational group at risk for occupational infectious diseases. These included bloodborne pathogens, such as hepatitis B virus (HBV), hepatitis C virus (HCV), and human immunodeficiency virus (HIV). Health care workers were also vulnerable to airborne diseases such as M. tuberculosis (TB), a risk shared by social service workers and corrections personnel. Another major concern was allergy to natural rubber latex gloves, a frequently used form of personal protective equipment against infectious diseases. More study was needed on the implementation and effectiveness of both the OSHA standard for bloodborne pathogens and the *Centers for Disease Control and Prevention (CDC) Guidelines for Preventing the Transmission of Mycobacterium Tuberculosis in Health Care Facilities*. Acute respiratory illness and vaccine-preventable illnesses were also identified as important issues for further study.

A DECADE OF PROGRESS

NIOSH and its partners have achieved many successes in the area of infectious diseases, particularly with bloodborne pathogens. NIOSH produced or contributed to recommendations for the use of sharps containers, an advisory on use of glass capillary tubes, and an Alert on needlestick injuries, titled *Preventing Needlestick Injuries in Health Care Settings*. In addition, NIOSH helped develop the 2001 CDC HIV prevention strategic plan and provided funding to the National Clinicians' Post-Exposure Prophylaxis Hotline (PEPline). This hotline provides free advice to clinicians treating workers who have been exposed to blood and other potentially infectious body fluids.

NIOSH also supported surveillance and epidemiology projects such as the CDC National Surveillance System for Healthcare Personnel (NaSH) and EPINet, a project developed by the International Health Care Worker Safety Center at the University of Virginia. These projects report how and why needlestick and other sharps injuries occur. Data from these projects, as well as NIOSH testimony, contributed to the Needlestick Safety and Prevention Act of 2001 and the subsequent OSHA-revised Bloodborne Pathogens Standard 1910.1030. This revised standard requires employers to select safer medical devices such as needle-less systems and sharps with injury protections and to involve employees in the selection process. It also mandates the use of newer, safer technologies as they become available.

Improved needle devices were not the only safety advance to occur during NORA. During the original NORA stakeholder meetings, partners identified allergies caused by natural latex rubber (NLR) gloves as an important problem. The use of NRL gloves skyrocketed in the early 1990s as standard precautions to prevent transmission of bloodborne pathogens were widely adopted. This dramatic increase in the production and use of NRL gloves was associated with an epidemic of NRL allergy in health care workers. Health problems included asthma, rhinitis, hives, and life-threatening allergic reactions. Use of cornstarch powder in gloves played a major role in this problem, because it carried the latex allergen in the air, exposing not only the users of the gloves but also bystanders.

In 1997, NIOSH responded to the emerging crisis among workers using NRL gloves with an Alert, titled *Preventing Allergic Reactions to Natural Rubber Latex in the Workplace*. This Alert strongly advocated for reducing exposures by using NRL gloves only when needed and, if their use could not be avoided, using non-powdered NRL gloves. The Alert also advised workers with NRL allergies to use gloves made of materials other than NRL and recommended that their co-workers avoid the use of powdered gloves to prevent bystander exposure. Total avoidance of NRL devices by both workers and co-workers was recommended as a final measure. Although controversial at the time, these recommendations were eventually widely accepted. The use of powdered NRL gloves has decreased in part because of the NIOSH Alert as more facilities adopt non-powdered NRL or non-latex options. The problem of NRL allergy in health care workers has decreased markedly and is considered a major public health success story.

TB was another important problem addressed by NORA. The emergence of HIV in the 1980s and 1990s was associated with a resurgence of TB as a public health threat in the United States. HIV-infected individuals were developing active, contagious pulmonary TB, resulting in numerous outbreaks of TB in health care settings. Outbreaks of multidrug-resistant TB were an especially important concern. In 1994, the CDC and a range of partners including NIOSH responded to this emerging problem by publishing the *CDC Guidelines for Preventing the Transmission of Mycobacterium Tuberculosis in Health-Care Facilities*. The guidelines were an aggressive collection of recommendations for administrative and environmental controls, including the use of respirators and medical screening. After NORA was established in 1996, efforts focused on implementing the CDC Guidelines. NIOSH contributed in the areas of environmental controls and use of respirators. Implementation has been a great success. TB outbreaks and rates of tuberculin skin test conversion among health care workers have been markedly reduced.

FUTURE DIRECTIONS

Future research priorities must focus both on general issues relevant to many infectious diseases and occupational groups and on priorities specific to particular diseases and occupational groups. Surveillance is an example of an important general priority. There is little or no organized surveillance to determine the workplace association of most infectious diseases, particularly those with little individual morbidity or mortality such as scabies or upper respiratory tract infections. Another surveillance issue is to evaluate the performance and applications of new, automated surveillance systems intended to identify outbreaks due to emerging infectious diseases or bioterrorism.

Other important research priorities will be in the areas of exposure reduction through environmental controls, personal protective equipment, and decontamination. Optimal technology and application of environmental controls such as ultraviolet germicidal irradiation, personal protection with respirators, and decontamination will be important areas of study. Other general research priorities include global infectious diseases, prevention effectiveness research, and needs of immune-compromised workers.

Many of the previous research priorities from the first decade of NORA will continue to be relevant. Bloodborne pathogens will remain a key research priority, not only for health care workers but also for diverse occupations including body piercers, tattoo artists, waste haulers and handlers, waste water workers, funeral service workers, and many others. The development and assessment of new control technologies to reduce needlestick and sharps injuries remain an important area. Should appropriate vaccines for agents such as HIV and HCV become available, rapid assessment of their effectiveness in the workplace will be a very high priority. Another high priority is determining the appropriate regimen and timing of anti-viral therapy for early (acute) HCV infection after occupational exposures.

Other priorities include research demonstrating the effectiveness of new in vitro tests for latent TB infection in the workplace and research aimed to improve adherence to treatment for latent TB infection. Additional issues specific to biowarfare agents include testing new technologies for automated surveillance and real-time exposure detection, as well as developing effective emergency preparedness systems. Finally, upper respiratory tract infections are the most frequent illnesses causing absenteeism from work and have tremendous economic impact despite their usually benign outcome. Developing effective prevention in this area could have important economic and health benefits.

Musculoskeletal Disorders (includes Low Back Disorders)

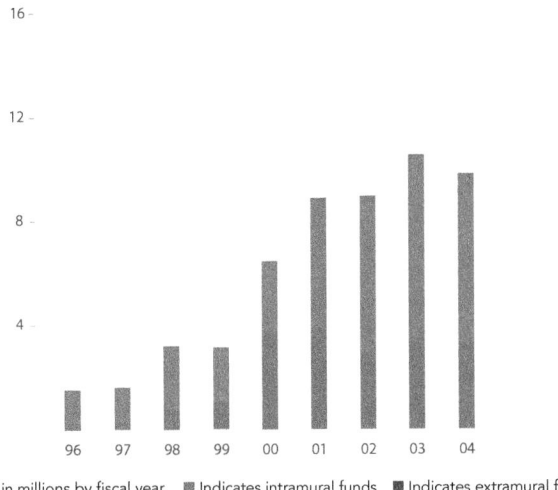

Funding in millions by fiscal year. ■ Indicates intramural funds ■ Indicates extramural funds

Musculoskeletal disorders (MSDs) affect the muscles, nerves, tendons, joints, cartilage, and spinal discs. They include ailments such as low back pain, shoulder disorders, tendonitis, and carpal tunnel syndrome. Excessive physical work demands, such as extreme muscle force, repetition, awkward postures, or fast movements, are known risk factors for these disorders. MSDs, however, are extremely complex. Individual factors, physical requirements, workplace organization, and psychosocial factors have all been associated with increased risk. The scope and toll of these disorders are enormous. In 2001, the Bureau of Labor Statistics (BLS) estimated that there were 522,528 MSD cases, with more than 43% of the cases involving more than 20 days away from work, a significant burden for both workers and their employers.

The primary focus of the team was to prevent work-related MSDs by:

> Evaluating the current status of scientific research relating to the prevention of work-related MSDs;
>
> Identifying gaps in the research base;
>
> Prioritizing and highlighting future research needs; and
>
> Facilitating research through development of partnerships with other government agencies and groups.

A DECADE OF PROGRESS

In 2001, the team published the *National Occupational Research Agenda for Musculoskeletal Disorders,* a document based on a series of regional focus groups with practitioners and academicians who helped identify significant research gaps and the following needs:

> Improving communication between those who conduct and those who apply research;
>
> Improving access to industrial sites to conduct research;
>
> Incorporating greater management and labor involvement in the research process; and
>
> Improving the dissemination of research results.

Partnerships involving government agencies, university researchers, private industry, and labor unions would be critical to bridging communication gaps, developing efficient research strategies, and improving the dissemination of information. NIOSH recently partnered with the National Institute of Arthritis and Musculoskeletal and Skin Disorders (NIAMS) to publish a request for extramural grant applications specifically directed at areas contained in the NORA research agenda. In 2003, the team partnered with Ohio State University to sponsor a national meeting, titled "State-of-the-Art Research Symposium (STARS): Perspectives on Musculoskeletal Disorder Causation and Control." The STARS conference provided a forum for experts from a diverse range of disciplines to present and discuss state-of-the-art knowledge relative to MSD causation and control. The range of topics included MSD epidemiology and economics, loading biomechanics, traditional and non-traditional pain tolerance, individual and genetic factors, psychosocial and organizational factors, and primary and secondary interventions. In 2004, a special issue of the *Journal of Electromyography and Kinesiology* published papers from many of the presenters at the conference. More recently, in 2005, the team partnered with the Washington State Department of Labor and Industries and the Northwest Center for Occupational Health and Safety to sponsor a national meeting, titled *"The Changing Nature of Musculoskeletal Disorder Risk: The Effect of Obesity and Aging in the American Workplace."* This conference focused on how work and individual factors interact to reduce risk, accommodation in the workplace, and methods to transfer research results into workplace practice. A series of research gaps were identified and will be published in the near future.

FUTURE DIRECTIONS

In the past, heavy physical demands have been the hallmark of work. Today's workplaces, however, are rapidly changing. Workers now move between work cells where they perform a variety of tasks, rather than stand in a traditional assembly line. The service sector is growing and often involves tasks conducted in a variety of non-conventional environments. E-commerce has sparked a vast increase in distribution center jobs. Computer usage is dramatically increasing. Researchers must examine risk factors and prevention strategies for these changing work environments.

Much of our understanding of MSDs has been gained through research of manufacturing workplaces and may not apply to office environments. We need to better understand, for example, the interaction between low-level static exertions and the mental demands experienced by computer users. Studies on cadavers and recent studies on animals have improved our understanding of how tissues respond to repetitive, forceful loading, but future research must describe this tolerance in healthy as well as compromised living human populations. Better surveillance also is needed to appreciate the magnitude of risk associated with shoulder loading in the workplace.

Interactions among risk factors will be an important research topic. Although we are beginning to understand how major risk factors affect human tissue, their interaction is virtually unexplored. Improved preventive strategies will depend on better understanding the links between biomechanical loading, soft tissue tolerance, and psychosocial stressors. As people live longer, and the average age of the U.S. workforce increases, the impact of aging on work-related loading, tolerance, psychosocial stress, and their interactions must be investigated. As the U.S. population becomes heavier, additional load is placed on the musculoskeletal system. The implications for workplace design and wellness have not been sufficiently explored.

The role of workplace factors in the development of fibromyalgia has been virtually unexplored, despite the fact that many symptoms of work-related MSDs resemble myofascial pain. Research efforts must focus on how low-level sustained or repetitive exertions, prevalent in the workplace, may influence muscle recruitment patterns, resulting in soft tissue disruption, pain, and dysfunction.

Research involving the risk of secondary injury associated with return-to-work is sparse. Integrating biomechanical exposures, soft tissue pathomechanics, and psychosocial factors into laboratory, epidemiological, and intervention studies is needed. Such studies will help determine the risk of injury to an individual who returns to work while recovering from an MSD.

Finally, alternative research designs are needed to more precisely assess the impact of interventions on the risk of workplace musculoskeletal injury. Most research has focused on the causal relationship between work and MSDs. Intervention effectiveness research can be achieved through standardizing research metrics and designs. Efforts must include randomized trials, wherever possible, and a quasi-experimental study design with control groups, at a minimum.

Reproductive Health Research
(formerly Fertility and Pregnancy Abnormalities)

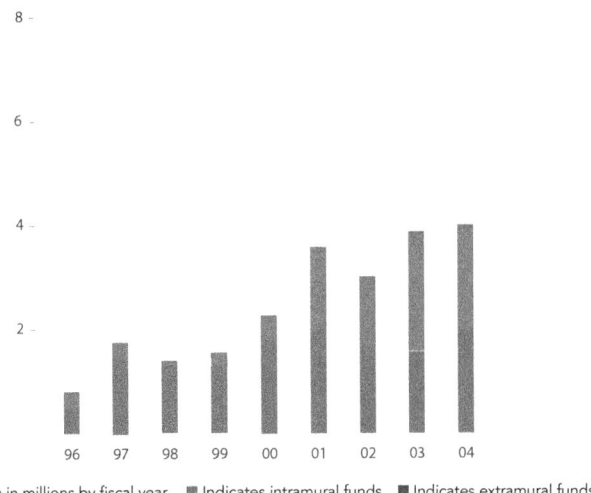

Funding in millions by fiscal year.   ■ Indicates intramural funds   ■ Indicates extramural funds

The initial challenge to this research is to effectively study the many toxicants, physical agents, and biomechanical and psychosocial stressors that may constitute reproductive hazards in the workplace. Although the main objective of researchers and clinicians is prevention of recognized adverse reproductive outcomes, there is growing need to examine chronic health outcomes potentially affected by reproductive toxicants.

A DECADE OF PROGRESS

Over the past decade, the NORA Reproductive Health Research Team has focused on improving coordination and feedback among disciplines and federal agencies, including members engaged in laboratory research, epidemiology, risk communication, and public health. The team's achievements have focused on prioritizing reproductive toxicants for further research, promoting the study of these toxicants, and promoting occupational exposure assessment in surveillance studies. The team established a National Occupational Reproductive Health Research Agenda to recommend future research directions to reduce the incidence of adverse reproductive health outcomes. This work can be accomplished with an interdisciplinary research program that identifies reproductive hazards, their toxic effect on the body, and target populations. Elements of this agenda include:

- New technologies and methodologies will help researchers do the following:
  - Understand toxicant mechanisms.
  - Identify populations at risk.
  - Evaluate reproductive/developmental hazards.
- Prioritize research needs:
  - Prioritize toxicology studies based on chemical structure and volume of use.
  - Prioritize field studies based on toxicological studies and human exposure information.
- Increase surveillance activities:
  - Evaluate occupational exposure data from existing surveillance systems.
  - Expand birth defects surveillance systems.
  - Add reproductive biomarkers and semen characteristics to national surveys.

Assess gene-environment interactions and toxicant mixtures in new studies, where appropriate.

Communicate non-technical research results to policy makers and the affected public.

Improve communication among researchers to bridge interdisciplinary gaps.

Approximately 84,000 chemical compounds are in the workplace, with 2,000 new chemicals introduced each year. Only about 4,000 of these chemicals have been evaluated for reproductive toxicity. A NORA-sponsored expert panel prioritized chemical reproductive toxicants identified by the National Toxicology Program (NTP) to encourage health studies of high-priority toxicants. Many more chemicals remain to be studied, and future priorities are likely to be affected by improved exposure information coming from the biomonitoring efforts of the CDC.

Priority toxicants have been the focus of new studies initiated both inside and outside of NIOSH. Federal spending for occupational reproductive health research increased. Team members have authored several grant announcements with other federal agencies for human occupational reproductive studies, with several grants awarded under each announcement and are the basis of additional grant announcements. NIOSH is studying several prioritized toxicants with field studies, exposure assessment, and laboratory biomonitoring.

To promote the study of high-priority reproductive toxicants, the team has established a partnership with the NTP Center for the Evaluation of Risks to Human Reproduction (CERHR). The Center provides scientifically based, uniform assessments of the potential for adverse effects on reproduction and development caused by agents to which humans may be exposed, which are summarized in terms that can be understood by those who are not scientifically trained.

In the 1990s, new scientific evidence documented widespread contamination of oncology clinics and pharmacies with anti-cancer hazardous drugs. Although safe-handling guidelines were published, research showed that there was poor adherence to recommended standards of safe professional practice. In light of the severity of the hazard, the potentially high number of workers affected, and the availability of interventions with high likelihood of success, the NORA Reproductive Health Research and Control Technologies Teams formed a multidisciplinary Hazardous Drug Safety Working Group.

The contribution of parental occupational exposures to birth defects has not been adequately studied, even though thousands of chemicals are used in the workplace by individuals of reproductive age. The CDC National Center on Birth Defects and Developmental Disabilities (NCBDDD) established Centers for Birth Defects Research and Prevention in several states to participate in the National Birth Defects Prevention Study (NBDPS), the largest case-control study of birth defects ever undertaken. NIOSH scientists are collaborating with the NCBDDD and the National Cancer Institute (NCI) to conduct an occupational exposure assessment using parental occupational information collected as part of the NBDPS.

Effectively studying the many exposures that are possible reproductive hazards in the workplace is challenging. The team has worked to improve occupational reproductive research by prioritizing reproductive toxicants for further study; promoting analysis of occupational exposure assessment in reproductive health surveillance; facilitating collaboration with biologists and toxicologists; promoting quality exposure assessment in field studies; and encouraging the design and conduct of priority occupational reproductive studies.

FUTURE DIRECTIONS

While continuing to expand and focus occupational reproductive health research, improving communication across disciplines, and facilitating joint efforts among different agencies, the team will base new activities on a broadening in research focus and new ways of thinking about exposures and reproductive outcomes.

> Rethinking outcomes and exposures. The changing nature of work and the work environment as well as emerging technologies in reproductive biology and exposure assessment are leading researchers to rethink approaches to studying "known" exposures and traditional reproductive health outcomes.
>
> Mechanistic research and tools for basic science. In order to make faster progress in analyzing the backlog of untested chemicals, a system for prioritizing them for the definitive testing that is done in laboratory rodent models is needed. There are a number of possible ways to screen chemicals that mimic one or more critical biological events that occur in humans and animals as part of the reproductive process. Continuing advances in our understanding of the underlying molecular control are making it possible to design structure-activity relationship programs and high-throughput screens that may be useful for prioritizing compounds based on presumed mechanisms of action and potency.
>
> Communication. The team is collaborating with the Hazardous Drug Safety Working Group to update written instructions and label warnings for certain hazardous drugs. The NIOSH Reproductive Health Research Team is also interested in finding ways to improve the quality of Material Safety Data Sheets (MSDSs), with special interest in improving the quality of reproductive health information. The team

conducted a session on MSDS Communication at the 2005 Society of Toxicology Meeting to help improve reproductive hazard communication.

Research to Practice (r2p) in Occupational Reproductive Health Research. Much of occupational reproductive research is etiologic, and this research has traditionally affected workers through the application of those findings to the regulatory process, which can be an extended process. However, a vibrant example of r2p implementation is the NIOSH Alert titled *Safe Handling of Hazardous Drugs in Healthcare*, which NORA's Hazardous Drug Safety Working Group sponsored. The Alert describes the unsafe handling of hazardous drugs in health care settings, an instance in which exposure opportunity is unregulated and the hazard is high.

Traumatic Injuries

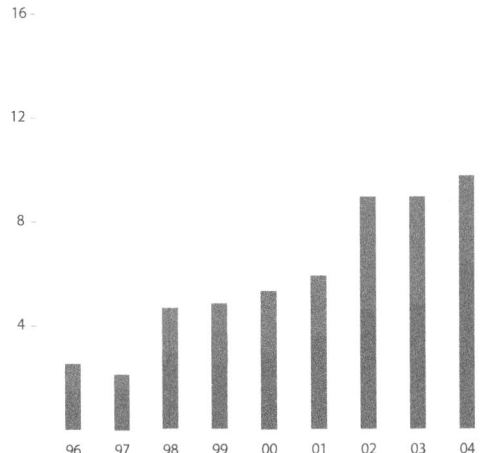

Funding in millions by fiscal year. ■ Indicates intramural funds ■ Indicates extramural funds

As the NORA Traumatic Injury Team explained in 1998, "Injury exacts a huge toll in U.S. workplaces—on an average day, 16 workers are killed and over 17,000 workers are injured. The associated economic costs are high—about $121 billion per year."

There has been progress in injury reduction as evidenced by the decades-long downward trend in the overall occurrence of traumatic occupational injuries and deaths in the U.S., despite the growth of the working population. Recent BLS statistics documented 5,559 deaths in 2003, yielding a fatality rate of 4 deaths per 100,000 workers. The BLS also reports non-fatal occupational injuries declined from 6.8 million in 1994 to 4.4 million in 2003. The toll of the injuries and deaths, however, continues to be unacceptably high, placing an enormous burden on workers, their families, employers, and the U.S. economy.

A DECADE OF PROGRESS

The NORA Traumatic Injury Team has worked to ease this burden by increasing research funding and by guiding the safety and health community to focus on data-driven prevention priorities. The team's most notable achievement occurred in 1998 with the publication of *Traumatic Occupational Injury Research Needs and Priorities: A Report by the NORA Traumatic Injury Team*. This NIOSH document emphasizes using the public health model for planning and prioritizing occupational injury research and prevention efforts. It also promotes forming new collaborations and partnerships to develop specific strategies that actually prevent these injuries and deaths.

Achieving this goal will require new interventions that are both effective and appropriate for the workplace. Advances have been made including the development of passive controls; the use of models and simulations in research; and the use of biomechanics and anthropometry to study injury risks and identify prevention options.

The transfer and translation of new knowledge and technologies are critical, and the rise of the Internet has offered this possibility by greatly increasing the flow of information. Challenges remain, however, in the areas of information management, organization, accessibility, and interface design. Addressing these areas will be crucial for tailoring relevant and timely risk information for specific audiences.

The past nine years have seen tremendous growth in federal funding for traumatic injury research and an increased professional dialogue. In their 1998 report, the team offered academic institutions and research foundations an outline of research gaps and needs. NORA funds have also supported conferences and workshops addressing topics such as truck driver safety and occupational violence, sparking innovative programs and partnerships. In 2002, Congress allocated funds to NIOSH to fund an initiative on occupational violence, and in 2004,

NIOSH sponsored a conference, titled "Partnering in Workplace Violence Prevention: Translating Research to Practice," to stimulate strategic research and interventions.

The team also helped to form a number of other partnerships, including working with BJC HealthCare Systems, Inc, to evaluate a back injury prevention program. Members of the NORA Traumatic Injury Team assisted in planning the first three National Occupational Injury Research Symposia (NOIRS), which were conducted in 1997, 2000, and 2003. The team organized and sponsored a special symposium at the National Safety Council Congress in 2001, titled "Making Science Work for You: A Symposium for Safety Professionals," and explored the use of company-level surveillance approaches.

The gains in knowledge about occupational injury risks and prevention have undoubtedly increased the growth of evidence-based prevention, although the full impact of NORA is difficult to gauge at this point. In the meantime, additional methods of evaluation are needed. One method is to examine changes in the number of injury research papers published in the open literature. The team is exploring other potential measures, such as the number of occupational injury research and prevention academic programs as well as the number of presentations, posters, sessions, and/or tracks at major health and safety conferences. The NORA Traumatic Injury Team also is gathering information about research and prevention efforts that have produced advances in addressing the recommendations from the 1998 NORA report.

FUTURE DIRECTIONS

While progress has been encouraging, gaps still remain. Improvements are needed in the completeness, consistency, and accuracy of company-level surveillance systems, which are vital to developing more data-driven prevention efforts. Finally, surveillance of system failures or property damage events also may identify potential injury risks and prevention opportunities.

Research efforts also must expand with multi-discipline and multi-sector collaborations needed to advance the science of injury prevention, with additional work to assess the value of prevention strategies. The new models, tools, and guidelines needed for safety practitioners in the field will require better dissemination, communication, marketing, and technology transfer. In addition, this research should continue to be data-driven, and new methods of data collection, analysis, and information dissemination need to be developed.

The second decade of NORA, however, will require new areas of emphasis, including:

> Increasing the implementation and evaluation of traumatic occupational injury prevention programs. Evaluation should consider the full impact on the injury experience including its impact on other workplace measures (such as productivity and product quality) and the cost of injury to the worker, his family, and employer. These many facets of injury should then be compared to the cost and effectiveness of prevention efforts.

> Focusing on research to practice, or the application of research findings to workplace prevention efforts. Continued progress in reducing workplace-related injuries and deaths will require sharing of knowledge, methods, technologies, products, and practices for a wide range of audiences.

Focusing on some high-risk work activities that have not received adequate attention, such as occupational motor-vehicle, machinery, and workplace violence injuries and deaths.

Increasing integration and collaboration to create a multi-sector and multi-disciplined research effort.

Increasing communication between disciplines engaged in traumatic occupational injury research and prevention.

Developing more practical tools for companies to assess and address injuries following a "data-driven" model.

Increasing the marketing of NORA, traumatic injury research, and the public health approach in general.

Increasing the visibility of health and safety sciences as useful approaches to occupational injury research and prevention.

Expanding existing methods and developing new efforts to evaluate the impact of NORA-sponsored traumatic injury research.

Continuing to improve surveillance and reporting methods, especially for non-fatal events.

Exploring opportunities to include injury prevention concepts into the educational curricula at the earliest possible levels, including elementary school and pre-school programs.

Increasing inclusive research sensitive to worker demographics.

WORK ENVIRONMENT AND WORKFORCE

Emerging Technologies

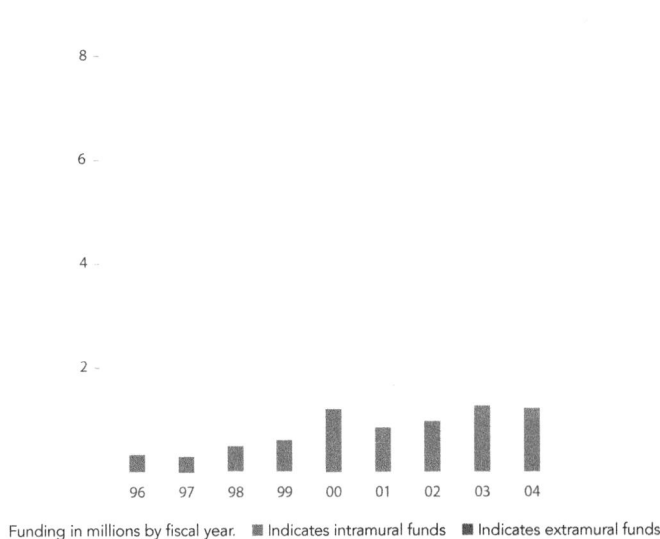

Funding in millions by fiscal year. ■ Indicates intramural funds ■ Indicates extramural funds

Today, almost everyone owns a cellular telephone. Cellular phones were "a science-based technology that created a new industry or radically transformed an existing one." This is the definition of an "emerging technology." The societal and industrial consequences of such technologies are often positive. However, since their development outpaces the understanding of their implications, they may pose new, unanticipated hazards. The cellular phone was implicated as a causative agent in human brain cancers. As a result, millions of research dollars were expended pursuing an answer. While debate ensued, research was conducted, but individuals continued to be exposed. Should exposure to cellular phones prove to be linked to human brain cancer, costs will be incalculable.

Formed in 1996, the NORA Emerging Technologies Team knew that a situation similar to the cellular phone story could unfold in occupational safety and health. Charged with protecting workers, the team faced the conundrum of designing prevention strategies for something that has not yet happened, is unanticipated, and absent of noticeable consequences. The team recognized that a new paradigm that moved from controlling identified hazards to anticipating, eliminating, or controlling the hazard before causing harm was imperative. Surveillance was absolutely essential in moving from a passive to an anticipatory mode. Predictive capacities for evaluating hazards would be responsive to rapid transformations occurring during the design of new technologies. The new paradigm would overcome the litigious and time-consuming delays in current risk assessments, and would recognize both the benefits and negative effects of emerging technologies. Finally, a proactive design for emerging technologies must consider how to eliminate hazards rather than just control them.

A DECADE OF PROGRESS

During the decade, the team refined its mission to (1) anticipate the potential occupational risks of new workplace processes, equipment, materials, and work practices; (2) assess the benefits of new technologies that can improve occupational safety and health; and (3) identify the needed industrial changes that have inputs, processes, and products that would be inherently safer for workers without compromising or transferring problems to the environment.

The team focused on anticipating hazards. Recognizing that there were no methods to anticipate, mitigate, or eliminate the potential hazards of emerging technologies, they created a research agenda and published it in a report, entitled *Emerging Technologies and the Safety and Health of Working People*. Centered on four knowledge gaps, the agenda presented for the first time an operational paradigm for occupational safety and health concerns related to emerging technologies. The gaps include: (1) Identification and Surveillance of Emerging Technologies, (2) Anticipating the Impact of Emerging Technologies, (3) Achieving Inherently Safer Design, and (4) Applying Spiral Development.

The team has been actively involved with many workshops. In 2000, they co-sponsored a workshop on the application of emerging technologies to ergonomics in conjunction with an American Society of Mechanical Engineers meeting. In 2001, the team co-sponsored the National Aeronautics and Space Administration (NASA) conference "Human Systems 2001: Exploring the Human Frontier." In 2004, they co-sponsored an international symposium regarding nanotechnology, and in 2005, NIOSH sponsored a conference, "Nanotechnology and the Safety and Health of Working People," at the University of Minnesota.

FUTURE DIRECTIONS

While progress has been steady, concerns remain. The four gaps provide a framework for future directions.

The first gap is to identify and prioritize which technologies require research. An academic team is evaluating the emerging technology literature. Methods for establishing minimum data requirements are necessary for effective early screening. Increased communication between occupational safety and health professionals and those developing new technologies is critical. Conferences, training, forums, and awareness building will aid in integrating diverse perspectives and in developing safer technologies.

Once a promising technology has been identified, a prospective analysis addresses the second gap by determining whether the new technology is safer and provides greater benefit than currently employed technologies. Prospective analysis also requires improvement, but when employed at each stage in the development of a technology, it successfully identifies research gaps so that critical needs can be addressed. Researchers require methods to test technologies at developmental stages for potential hazards, possibly resulting in redesign. Evaluation criteria and new analysis methods that include not only the hazards, but also the benefits of emerging technology, need to be developed. To illustrate the first two gaps, the team has identified nanotechnology as an emerging technology and a prospective analysis is currently in progress.

The third gap involves developing safer designs to identify alternatives to riskier technologies and to reduce or eliminate occupational safety and health problems. The chemical process industry has developed design principles for inherently safer technologies that can be explored in NORA sectors. Inherently safer designs need to inform applied science and engineering to help develop cleaner industrial processes. Research must consider worker safety and health when developing new technologies primarily designed to improve the environment. Achieving inherently safer designs requires new methods to compare alternative designs and to derive a better understanding of their receptivity in the workplace. Research is needed to recognize and overcome barriers in improving workplace conditions, while maintaining high-quality and innovative products and services.

The last gap focuses on creating an integrated process for adopting beneficial emerging technologies and avoiding safety and health problems. Such a process would integrate identification, analysis, and design methods to reduce or eliminate risks while maximizing the benefits of new technologies. A prevention approach necessitates the development of specific innovative technologies to control or eliminate perennial occupational illnesses or injuries. There is a need to investigate opportunities for using information technology or electronics and communications to monitor and inspect workplace programs. Research is necessary to expand the precautionary principle to U.S. workplaces.

The team recognized that partnerships are critical for identifying research associated with rapidly changing technologies. Research linking emerging technologies and occupational safety and health should be extended into specific occupations such as the construction, agricultural, mining, and service industries, as well as energy conservation, production, and storage.

Opportunities for collaborative research exist in several areas. Technologies developed to provide security have applications to protecting workers involved in homeland security. Scientific initiatives are opportunities for collaboration. Nanotechnology is an initiative authorized by the federal government and being developed in the private sector. National Science Foundation programs also offer opportunities for collaboration, particularly for partnerships between academia and the private sector. Environmental initiatives offer the opportunity for extending environmental concerns to include indoor and outdoor work environments.

As NORA embarks on its second decade, new areas requiring emphasis include:

- Increasing the implementation and evaluation of emerging technology programs in diverse occupational settings;
- Focusing on research to practice (r2p), or the translation and transfer of research to the workplace;
- Focusing on some high-risk work activities that have not received adequate attention;
- Increasing integration and collaboration;
- Increasing communication and collaboration among disciplines;
- Developing more practical tools for companies to assess and address emerging technology issues;
- Increasing the communication and marketing of emerging technology research;
- Increasing the visibility of the safety sciences as useful approaches to emerging technology research;
- Expanding existing methods and developing new efforts to evaluate the impact of NORA-sponsored emerging technology research;
- Continuing to improve surveillance and reporting methods, especially for non-fatal events; and
- Exploring opportunities to incorporate emerging technology concepts into educational curricula at the earliest possible levels.

Indoor Environment

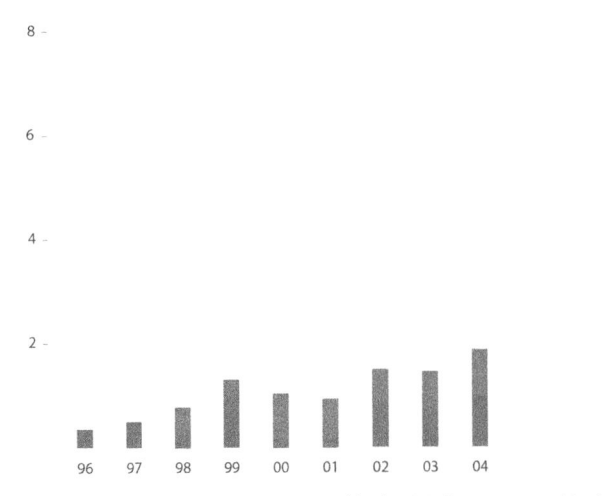

Funding in millions by fiscal year. ■ Indicates intramural funds ■ Indicates extramural funds

Almost 70% of U.S. workers are employed in non-industrial, non-agricultural indoor settings, referred to here as indoor work environments. Scientific studies have associated some indoor environmental conditions with increased risks of non-specific symptoms, respiratory disease, and impaired performance. The potential health and economic benefits of improving indoor work environments were largely unrecognized in the U.S. in the mid-1990s and remain so today. When the NORA Indoor Environment Team commenced in 1996, a national research effort was needed to establish strategic priorities to identify and implement health-protective features and practices in buildings.

A DECADE OF PROGRESS

The team published a white paper in the *American Journal of Public Health* in 2002. The white paper provided comprehensive estimates, based on previously published data, regarding the magnitude of health effects related to poor indoor air quality. The team estimated that modest improvements in indoor environments could prevent respiratory infections or exacerbations of asthma or allergies among 6 to 10 million workers annually. Improvements could also reduce frequently experienced building-related symptoms experienced by 8 to 30 million workers. The potential economic benefits from reducing these adverse health outcomes are estimated at billions of dollars for workers and employers annually.

The white paper also identified three interrelated categories of high-priority research needs. More research was needed to understand both the causes and prevention of building-related health effects. Conducting this research would require advances in the science and technology of indoor environments. Strategies to reduce barriers and increase incentives for health-protective building practices needed to be identified and evaluated. These recommendations were widely circulated. Reprints of the team's white paper were mailed to 3,000 national and international stakeholders in the fields of indoor environmental science and public health.

Presentations based on the research priorities outlined in the team's white paper also were made at a number of high-profile meetings. These meetings include the "132nd Annual Meeting of the American Public Health Association" in 2004; the Society for Occupational and Environmental Health's meeting on "Mold-Related Health Effects" in 2004; and the 2005 "Surgeon General's Workshop on Healthy Indoor Environment." In addition, the team conceived and sponsored, with additional support from the Harvard School of Public Health, a unique 2004 workshop, titled "Indoor Chemistry and Health," at the University of California, Santa Cruz, campus. This workshop assembled researchers for the first time to discuss "indoor chemistry." It generated hypotheses to test potential connections between indoor chemistry and human health and promoted interdisciplinary and international collaborations.

Approximately 70 participants from 8 countries met to discuss adverse health effects that might result from exposure to the products of reactions among indoor pollutants. Scientists from multiple disciplines including chemistry, toxicology, medicine, epidemiology, and public health addressed this complex subject.

The team also formed a number of partnerships, including the following successful collaborations:

> Team members contributed to a National Academy Press report on the Federal Facilities Council (FFC) workshop "Implementing Health Protective Features and Practices in Buildings." This workshop was based on one of the team white papers.
>
> An ongoing partnership with the National Center for Environmental Health (NCEH) resulted in several joint activities, such as NCEH purchase of state-of-the-art portable spirometers for the NORA-funded project "Work-Related Asthma in School and Office Buildings."
>
> In 2004, NCEH funded contracts for an international expert in indoor environmental studies of respiratory health and an expert in the clinical evaluation of upper respiratory disease to build capacity in NIOSH and NCEH staff. Discussions between NCEH and NIOSH yielded an agreement to collaborate to study indoor air quality and asthma in schools. The anticipated impact of such work will be guidance to state departments of education and school districts to determine the best use of their resources to provide healthful school environments.
>
> Finally, partnerships between the NORA Indoor Environment Team and a number of agencies including the Public Health Service, NCEH/ Agency for Toxic Substances and Disease Registry (ATSDR), NIH, EPA, and NASA, led to the highly successful "Surgeon General's

> Workshop on Healthy Indoor Environment," held in January 2005. This national partnership mirrors a larger global effort. In 2002, two members of the NORA Indoor Environment Team were invited to serve on a World Health Organization (WHO) expert panel addressing guidance for biological exposures in indoor environments.

While NORA has increased dialogue about indoor air research, it also has generated substantial original investigations. In 2000, NIOSH initiated the Research on Occupational Asthma Reduction (ROAR) program, which was made possible by NORA funding. The NORA ROAR project "Work-Related Asthma in School and Office Buildings" has documented in the peer-reviewed literature excess physician-diagnosed asthma, decreases in quality of life outcomes, increased use of sick leave, and validation of symptom reports by objective medical tests in occupants of a large water-damaged office building. Staff on this NORA-funded project have developed a visual assessment tool for documenting water damage and mold in buildings. Ratings assigned by the tool have been shown to have a positive association with questionnaire data for occupant health effects. Measured exposures from field studies in offices and schools are providing strong evidence for using dust as a historical marker of building contamination. Several microbial components of dust have shown positive associations with occupant health effects. Anticipated impact from this project includes the development of simple, practical tools for the assessment of damp environments by facility managers to guide priority setting for building remediation.

The team's research agenda and increased NIOSH extramural funding have stimulated an increase in the number and rigor of indoor environment-related research proposals in the United States. NIOSH funded seven extramural projects between 1998 and 2003. The studies have examined diverse issues, ranging from ventilation to the health and socioeconomic consequences of non-specific building-related illness.

FUTURE DIRECTIONS

The team's work during the first decade of NORA has helped focus the attention of stakeholders and decision makers on health issues relating to indoor environments. Important next steps are to leverage existing resources for research through partnerships among government agencies and other organizations that share interests in public health, work environments, energy use, and business productivity. These groups must then expand resources available for research by developing common interests among organizations that share goals of improving the nation's health, productivity, and economic competitiveness.

The priority research aims as outlined in the team white paper remain pertinent to lead the way into the next decade of NORA. Specific goals for the future should include:

> Improved assessment of the economic consequences of unhealthful indoor environments;
>
> Attention to reducing building-related respiratory infections;
>
> Improved methods for measuring exposures;
>
> Development of biomarkers indicating exposure and health effects;
>
> Attention to mixed exposures (both biological and chemical) in the indoor environment in relation to adverse health effects; and
>
> Improved understanding of the effectiveness of indoor environment interventions in reducing exposures and alleviating associated health effects.

Mixed Exposures

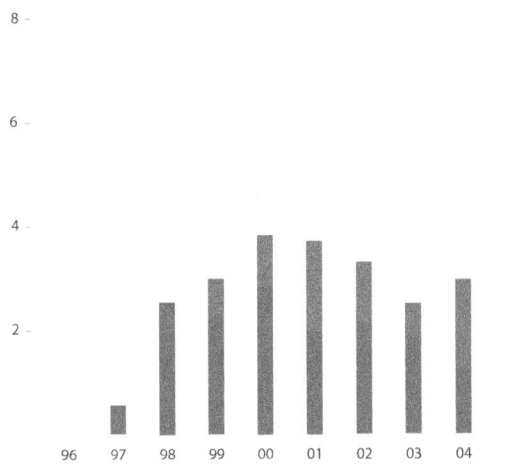

Funding in millions by fiscal year. ■ Indicates intramural funds ■ Indicates extramural funds

Workers are exposed to a wide variety of chemical, biological, and physical stressors. The interactions from these mixed exposures can increase the severity of their harmful effects. Exposure to both noise and the solvent toluene, for example, results in a higher risk of individual hearing loss than exposure to either stressor alone. Mixed exposures occur every day at the worksite and in the home, making them a multifaceted problem with a potentially enormous reach.

Support for this assertion is found in surveys such as the National Occupational Exposure Survey and the National Occupational Health Survey of Mining. These data indicate mixed exposures affect a large number of workers in a wide range of occupational settings, often with serious health outcomes. For example:

> More than 10 million workers are exposed to mixtures of fuels and combustion products that may cause cancer, chronic obstructive pulmonary disease, pulmonary function changes, chemical pneumonia, central nervous system effects, liver or kidney damage, and irritation of eyes, skin, or mucous membranes.
>
> Approximately 4.7 million workers are exposed to combinations of chemicals and noise, which may lead to hearing loss.
>
> More than 750,000 workers are exposed to mixtures of metal fumes and welding fumes, which can cause cancer, respiratory disease, metal fume fever, eye damage, and neurological impairment.

These statistics illustrate that the traditional one-chemical-at-a-time approach to occupational health is often not adequate. Risk assessments based on substance-by-substance or hazard-by-hazard approaches cannot evaluate additive or synergistic effects. The need to develop more advanced methods prompted stakeholders to designate mixed exposures as a NORA priority area.

A DECADE OF PROGRESS

Once a neglected research area, the study of mixed exposures has transformed from simple descriptive studies of binary mixtures to sophisticated investigations using new biological and computational technologies. The complete elucidation of the human genome, the related developments in genomics and proteomics, and the exponential growth of computational technologies now enable researchers to deal with the effects of mixed exposures on complex biological systems.

Developing a research agenda in response to these rapid advancements has been a major accomplishment of the NORA Mixed Exposures Team. *The Mixed Exposures Research Agenda*, published in 2004, articulates many of the issues involved with mixed exposures, recommends research strategies, and defines priorities that could lead to improved interventions for protecting workers.

Based on the reality of limited resources, the team identified several high priority research areas, including:

- Developing and implementing new surveillance methods to identify the number of exposed workers, the range of exposure concentrations, and their related health effects.

- Promoting collaboration between occupational health professionals and workers to rank and characterize mixed exposures. Such a strategy will also facilitate the dissemination of research findings.

- Conducting research to better understand how mixed exposures affect the body.

- Developing methods to understand and integrate experimental data from the molecular level and to extrapolate these data to whole body systems.

- Developing methods that can be used to measure and predict deviations.

- Developing and validating exposure-response models.

- Developing the concept of a virtual human through physiologically based pharmacokinetic simulation.

- Developing parameters for mechanistically based risk estimation and extrapolation models.

- Developing biosensors or measurement technologies that indicate whole mixture toxicity.

- Identifying, validating, and characterizing the health outcome for exposure and response biomarkers.

- Determining the effects of mixtures on engineering controls and personal protective equipment.

FUTURE DIRECTIONS

Addressing the Mixed Exposures Research Agenda will require new strategies to engage researchers, workers, and employers. Understanding the health effects of real-world mixed exposures will be mind-boggling unless systems are in place for clarifying research priorities within major occupational groups. Such a system should rank mixed exposures according to knowledge about their health effects and the likelihood of their occurrence. This structure will help create manageable priorities for research and worksite interventions and will necessitate collaboration between professional researchers, employers, workers, and the organizations that represent and train them.

A sector-specific approach will include traditional public health responses to study the multitude of potential outcomes resulting from mixed exposures. For example, mixed exposures may produce acute effects, chronic effects, or a combination of the two. These effects may be latent or active. Mixtures also may increase negative health effects, create unexpected health effects, or interact in the environment to generate new exposure risks. Surveillance, interventions and controls, and evaluation research are needed to identify and manage these risks.

The construction and mining industries illustrate the value of a sector-based public health approach. In construction, both old and new exposures may endanger a wide range of trades and occupations. Pipe fitters, for example, may be exposed to fumes from modern high-nickel alloy welding rods and to asbestos applied to pipes over a generation ago. Miners may be exposed to combinations of particulate matter released from diesel engines, irritant gases such as nitrogen dioxide, and asphyxiates like carbon monoxide. Every industry has unique combinations of stressors that can cause current or future occupational diseases. A sector-based public health approach will enable researchers to identify, study, and ultimately control the most urgent safety and health risks for specific industries.

Other research needs, however, will affect multiple sectors. Nanotechnology, for example, is an emerging and enabling technology that promises unprecedented advances in many diverse fields. This technology also has the potential to create new mixtures of chemicals, chemical forms, particle sizes, and routes of exposure. The potential for mixed exposures should be considered when evaluating practical approaches, such as control banding, for the safe manufacturing and handling of nanomaterials. This caution is especially important where insufficient information is available to apply traditional exposure-limit control strategies.

In addition, non-work-related exposures affect all sectors to various degrees. The use of alcohol, tobacco, insect repellents, cosmetics, or other chemicals may have significant interactions with workplace exposures. Individual susceptibility also adds to the complexity of exposures and their resulting biological responses. New approaches are needed to identify additive, synergistic, antagonistic, or potentiation effects from multiple exposures.

The *NORA Mixed Exposures Research Agenda* describes a variety of evaluation tools that can be used to assess the risk posed by mixed exposures in real-world settings. One approach is based on observed health effects and observed exposure risks. This approach ensures that protective technologies are not compromised by multiple simultaneous exposures, as in the case of an interfering agent reducing the service life of a respirator cartridge. Other evaluation tools and approaches will require additional research, such as the development of better tools for toxicity analysis, exposure-response modeling, and physiologically based modeling. Better guidelines are needed to assist occupational hygienists and other safety and health professionals.

Organization of Work

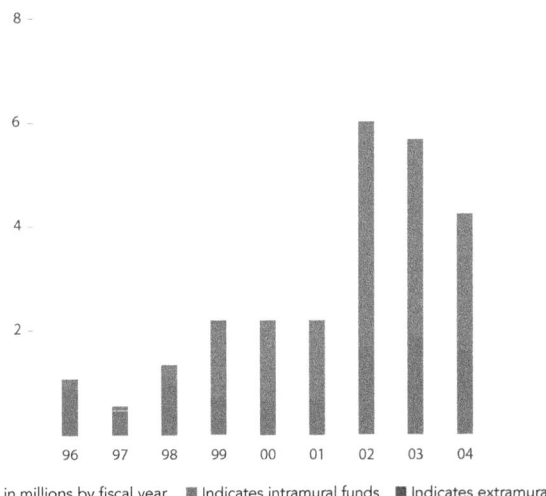

Funding in millions by fiscal year. ■ Indicates intramural funds ■ Indicates extramural funds

The organization of work refers to how jobs are designed and performed, to organizational practices that influence job design, and to broader economic, public policy, and other forces that encourage or enable new organizational practices. In the last two decades, U.S. companies have restructured and downsized their workforces in response to economic pressures. Many have increased their reliance on non-traditional employment practices and have adopted more flexible and lean production technologies. Concern exists that these trends may adversely affect job design, threatening worker safety and health.

In 1997, a diverse team of researchers and practitioners formed the NORA Organization of Work Team to review current knowledge, potential safety and health consequences, and possible prevention measures. The result was a comprehensive research agenda.

A DECADE OF PROGRESS

A review of the literature and input from stakeholders suggested numerous risks associated with changing organizational practices. Four priority areas of research and development were identified:

> Surveillance. Implement new and improved data collection efforts to better understand how the organization of work is changing, and how these changes may be affecting workplace exposures to risk factors for stress, illness, and injury.
>
> Etiologic research. Examine the health and safety effects of prominent trends in the organization of work that have arisen in recent years.
>
> Prevention research. Conduct intervention research investigating organizational practices and policies that may protect worker safety and health.
>
> Capacity building. Foster a stronger public health commitment, including steps to formalize and nurture organization of work as a distinctive field in occupational safety and health, development of training for research and practice in this field, and improved research-funding opportunities.

In 2003, subteams were organized to study (1) trends toward increasing work hours and implications for worker safety and health, and (2) disproportionate effects of the changing organizational practices on working women. Several intermediate outcomes are included among the products listed below.

The team undertook numerous initiatives to address needs for research and development to better understand and prevent hazards from the changing organization of work. Outputs include numerous publications and conferences that increased attention to this field and highlighted new research and knowledge.

Additionally, a leading peer review journal has committed to the 2005 publication of the organization of work subteam research agenda on long working hours together with other papers on this topic. Similarly, the publications arm of a major professional association has committed to the 2006 publication of an edited volume on organization of work and women's health that features the subteam research agenda on this topic.

The team also co-sponsored several public health conferences on organization of work:

> "Work, Stress, and Health: New Challenges in a Changing Workplace" (2003). With the American Psychological Association (APA).
>
> "Long Working Hours, Safety, and Health: Toward a National Research Agenda" (2004). With the University of Maryland and the Department of Justice.
>
> "The Way We Work and Its Impact on Our Health" (2004). With the University of California Centers for Occupational and Environmental Health in Northern California (University of California, Berkeley [UCB], University of California, San Francisco [UCSF], and University of California, Davis [UCD]) and in Southern California (University of California, Los Angeles [UCLA], and University of California, Irvine [UCI]).
>
> "4th International Conference on Work Environment and Cardiovascular Diseases" (2005). With UCI and UCLA. the Mt. Sinai School of Medicine, the Japan Association of Job Stress Research, and APA.
>
> "Work, Stress, and Health: Making a Difference in the Workplace" (2006). With APA, the National Institute of Justice, the Department of Labor, and the National Institute on Disability and Rehabilitation Research.

Efforts under NORA have led to substantial progress to better understand and prevent safety and health risks associated with the changing organization of work.

Progress on the research agenda:

> Improved surveillance. A 2002 follow-up to the 1969 to 1977 "quality of employment" worker surveys was conducted by the National Opinion Research Center (NORC) to review how job demands have changed. In 2003, NORC also administered a revised version of the 1991 and 1996/1997 National Organizations Survey of establishments to better understand trends in changing organizational practices. Finally, an organization of work module is being designed for the new National Exposures at Work Survey under development at NIOSH.
>
> Increased support for research. The number of NIOSH intramural projects classified as organization of work research has increased since 1996. Studies in the occupational safety and health literature that address organization of work issues are increasingly prevalent.
>
> Research capacity building. In 1996, NIOSH entered into a cooperative agreement with APA to foster graduate training in health, safety, and the organization of work. By 2002, start-up funding had been provided to 11 universities for graduate programs that blended training in psychology and occupational safety and health. Fifty-three students completed this course of study within the time frame of the cooperative agreement (1996 to 2002). By 2004, two universities had successfully competed for NIOSH training grants for sustained graduate training programs with this interdisciplinary focus.

Additionally, in 1999, the NIOSH Applied Psychology and Ergonomics Branch was reorganized and renamed the Organizational Science and Human Factors Branch to provide increased visibility and support for organization of work research within NIOSH.

FUTURE DIRECTIONS

Despite gains, safety and health knowledge is lagging, and challenges remain. But perhaps a more fundamental concern is that organization of work is not yet an established field of study. Numerous disciplines have contributed research, but there is little interface and important differences exist among them.

Stronger alliances and collaborations are needed. These relationships could lead to greater visibility, provide a foundation for the multidisciplinary training needed for research in this field, and provide a framework for systematic progression of research—all of which may, in turn, improve funding opportunities. Universities funded under the APA/NIOSH cooperative agreement to cultivate training programs in work organization and health are formalizing a new discipline, known as "Occupational Health Psychology," and supporting professional society that blends content from many fields that contribute to research on the organization of work.

Special Populations at Risk

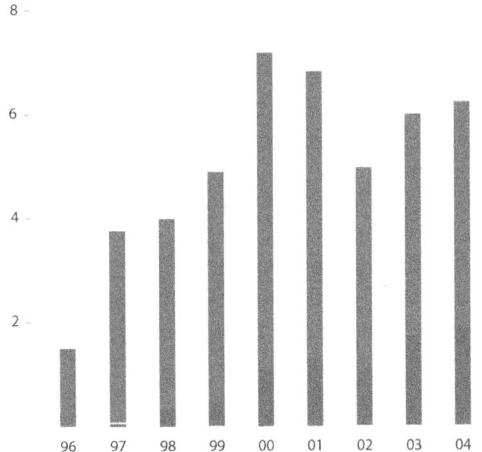

Funding in millions by fiscal year. ■ Indicates intramural funds ■ Indicates extramural funds

The U.S. workforce is transforming. Today's workforce contains many more women, older workers, and racial and ethnic minorities, and virtually all workforce expansion in the past decade has been through increased numbers of foreign-born workers. These groups each experience disparities in the burden of disease, disability, and death, due in part to their disproportionate employment in high-hazard industries and certain social, cultural, and political factors. Minority workers, for example, may encounter discriminatory employment practices. Immigrant workers may have literacy challenges that compromise the effectiveness of traditional training and labeling practices. Low-income workers may lack health insurance or access to health services. Older workers may face physiologic and cognitive challenges that have not been adequately researched. The NORA Special Populations at Risk Team identified and promoted research to better characterize these risks and to help develop effective intervention programs.

A DECADE OF PROGRESS

In 1999, the journal *Occupational Medicine: State of the Art Reviews* dedicated an entire issue to Special Populations. This volume described the magnitude of risk for these diverse worker populations and their major research needs. In addition to this broad cross-cutting research agenda, the team defined and promoted specific research needs for some of the highest-risk populations, including older workers, foreign-born workers, women, and low-income workers.

Through NORA, NIOSH has prioritized research on the health and safety of workers ages 55 and older, a group that will increase by 70% in the next 10 years (BLS). Several pioneering research studies are currently being completed. In 2004, the National Research Council released a report on the Health and Safety of Older Workers. The team also developed a document outlining the major research needs including improved surveillance, research, and interventions for older workers.

Several team projects have fostered collaboration with community-based organizations, which are perhaps best suited to respond to the unique linguistic, cultural, and legal barriers facing immigrant workers. NIOSH co-funded an RFA with the National Institute for Environmental Health Sciences, titled "Environmental Justice: Partnerships for Communication." This solicitation funded studies that engaged immigrant communities in identifying, defining, and developing solutions to safety and health problems. Since the initiation of this extramural funding program in 2003, eight research projects have studied workplace hazards to Asian and Hispanic immigrants employed in agriculture, poultry processing, construction, and the service sector. In 2004, community-based organizations joined health researchers, government officials, and advocates to discuss research gaps and challenges at the "Symposium on Immigrant Worker Safety and Health." The proceedings of this conference will be available in 2006. The team also supported the National Center for Farmworker Health to create a research track at their migrant stream forums, including special sessions at the annual meeting and an Internet listserv for migrant health research.

The team partnered with the National Institute for Child Health and Human Development (NICHD) to identify workplace practices and policies that impact the health of employees and their families and dependents in order to design effective workplace-based interventions. Collaboratively, the team and NICHD co-sponsored three scientific workshops and a 2005 RFA. The projects funded through this initiative will investigate how workplace policies and practices can be changed to improve the health and well-being of a diverse group of workers and their family members, including among supermarket workers, hotel hourly workers, low-income workers employed by small manufacturing facilities, and workers employed in high-tech industries.

In 2000, the team partnered with NIH to co-fund an RFA, titled "Health Disparities: Linking Biological and Behavioral Mechanisms with Social and Physical Environments." This program sought multidisciplinary research to better understand how social and physical environments lead to health disparities. The physical environment included physical, chemical, and biological exposure agents in the workplace and community. The social environment included individual and community-level characteristics, such as socioeconomic status. Through this solicitation, three 5-year occupational health projects were funded to study injuries, work stress, and musculoskeletal disorders among health care workers and rural female African American poultry processing workers.

Inside NIOSH, several new and continuing NORA-supported intramural initiatives have also promoted research concerning special populations at risk. Pilot research projects have identified data needs and intervention approaches for the following kinds of workers: young, minority, immigrant, older, and disabled. A project also has examined how occupational variables might better be integrated into NIH-financed health disparities research. The team co-sponsored a planning meeting with the Office of Behavioral and Social Sciences Research of the Office of the Director of NIH in April 2004 to obtain input from leading researchers in social epidemiology to better understand current practices in the measurement and analytic treatment of occupational variables in population-based health disparities research.

FUTURE DIRECTIONS

During the past decade, concerns related to health disparities in the U.S. have become increasingly important and have received greater attention, including heightened attention by the occupational health community. The team has contributed to the creation of a solid base of ongoing research that should lead to the development of innovative model health communication and prevention programs for underserved working populations. The various initiatives described above have all pointed to several cross-cutting themes that will be important to help guide future work in this priority area.

Although many working populations fall under the broad category of Special Populations at Risk, workers of low socioeconomic status or low income, whether they are any of the following: youth, older, immigrant, disabled, or minority, appear to be at greatest risk for health disparities. Future plans need to emphasize this subgroup of special working populations.

Future activities also should focus on research addressing the additional impact of race and ethnicity on the risk for occupational health disparities, including identifying the role of institutional forms of racism as well as other sources of discrimination resulting from bias based on race/ethnicity, language, or cultural or political factors, such as immigration status.

The use of community-based participatory research (CBPR) methods should continue to be a focus of research with special populations. Although occupational health research traditionally has focused on workplace-based research, it is often important to combine these approaches with community-based methods when working with special populations. CBPR promotes improved data collection on the social, cultural, and political factors influencing workplace safety and health and will require greater emphasis on qualitative research methods. CBPR should also focus on improved efficacy of occupational intervention programs aimed at special populations at risk.

Improved surveillance methods are needed to better track the frequency of occupational injury and illness among special populations. Many surveillance systems do not routinely collect data on race, ethnicity, or immigration status, making it difficult to track these variables. Under-reporting of events in special populations is probably high because of factors such as contingent and temporary work status and fear of potential job loss for reporting injuries.

The tremendous increase in the number of professionals interested in health disparities research provides opportunities for unparalleled collaborations with other fields of health research. Since many researchers have identified socioeconomic status as a key variable for their work, occupational health researchers can provide important insights into the development of accurate measurement tools for the work-related components of socioeconomic status.

RESEARCH TOOLS AND APPROACHES

Cancer Research Methods

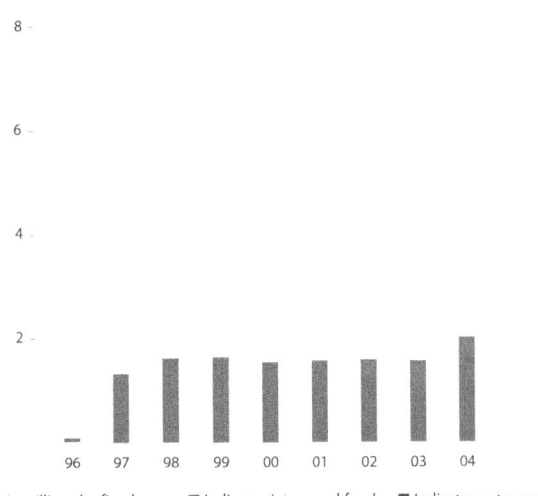

Funding in millions by fiscal year. ■ Indicates intramural funds ■ Indicates extramural funds

The workplace significantly contributes to the human cancer burden, exceeded only by cigarette smoking and diet. Each year in the U.S., approximately 600,000 deaths occur from cancer; between 4% and 10% of these deaths are estimated to be from workplace exposures. This figure, however, may be underestimated. Estimates of occupationally related cancers are based on cancer sites such as lung and bladder, which are recognized as having a substantial occupational component. As a result, these figures do not account for any additional contribution that workplace exposures play in cancers of other sites. In addition, the burden of recognized occupationally related cancer falls especially on workers in blue collar jobs in high-exposure industries—mining, construction manufacturing, and certain parts of the service sector.

Although in 1996 various countries had reduced workplace exposures to some carcinogens, many substances listed by the International Agency for Research on Cancer (IARC) as possibly (2A) or probably (2B) carcinogenic to humans had not been regulated as carcinogens by OSHA. Numerous occupations also demonstrated an elevated risk of cancer, but investigators could not definitely identify a causative agent. These occupations include painters, rubber workers, dry cleaners, welders, and workers who use printing processes. Finally, exposures to particulates, combustion products, and complex mixtures such as asphalt and metalworking fluids continued to provide toxicological and epidemiological challenges. Because of these factors, it was critical to assess how the development and application of new methods could enhance the study of occupational cancer and contribute to its prevention.

A DECADE OF PROGRESS

The team broadly defined cancer research methods to include the range of methods, tools, approaches, and strategies that enable occupational cancer research. The team supported a comprehensive approach, including computer modeling, in vitro and in-vivo testing, epidemiologic studies, and exposure characterization.

The first task that the team undertook was to develop a research agenda that addressed four broad areas: identification of occupational carcinogens, design of epidemiologic studies, risk assessment, and primary and secondary prevention. The result of this effort was a paper, entitled *Priorities for Development of Research Methods in Occupational Cancer.* The paper included a review of research gaps and needs, and presented approximately 30 recommendations to address them.

A consensus emerged from the discussion of the research agenda that there was a need to focus on how to apply new biotechnologies to the study of occupational cancer. An international workshop held in May 2002 featured 80 researchers who study worker populations or develop and validate new biotechnologies. The workshop focused on three broad topics: (1) the challenge of applying new technologies to the study of occupational cancer,

(2) markers of early biologic effect, and (3) the application of genetic biomarkers to human studies.

A summary of the workshop *Applying New Biotechnologies to the Study of Occupational Cancer* was published in 2004. Notable findings included:

> The capability of new biotechnologies to group chemicals with similar global gene expression profiles has the potential to provide an early warning system for suspected carcinogens before they are introduced into commerce. The challenge will be to identify the degree of similarity required to predict carcinogenicity and to distinguish pathogenic patterns from homeostatic ones.
>
> Gene expression patterns may be used in epidemiologic studies as surrogate endpoints for cancer. Attention to basic epidemiologic principles of design and analysis is still important to guard against biases and irreproducible results. To enhance risk assessment, expression patterns need to demonstrate comparability across species for extrapolation purposes, and need to be robust at different doses for dose-response predictions. Before these technologies are used in humans, the ethical, legal, and social issues should be addressed along with the scientific issues.

Various projects were developed based on the research agenda. These included an evaluation program to determine research needs of IARC Group 2A, 2B, and 3 carcinogens, identification of characteristics for a fellowship program to attract and nurture new investigators, and various biomarker guidance and database efforts.

FUTURE DIRECTIONS

In the next decade, it will be necessary to identify what new high-throughput biotechnologies—particularly genomics, transcriptomics, and proteomics—can provide for occupational cancer research. The products of these new biotechnologies may be considered as new biological markers. A large number of gene variants, transcripts, or proteins can now be assessed in a very short time. These products depict an increased level of complexity because they represent a more detailed and system-wide data set.

These biotechnologies also offer the opportunity to understand how genetic polymorphisms contribute to the risk of occupationally related cancer and to identify some carcinogens whose effects may be apparent only in genetically susceptible populations. It is unlikely that a single metabolic polymorphism is a major component in occupational cancer risks; rather, genetic polymorphisms are involved in many mechanisms that act in concert or conflict and need consideration simultaneously. The high-throughput biotechnologies will allow these complex interactions to be explored for the first time. Before their utility is assured in occupational cancer research, however, a number of technical and scientific issues need attention. These issues include standardization of techniques, decisions about scales and outliers, and methods for comparing platforms and understanding the meaning of various perturbations of array patterns in relation to exposure, effect modification, or disease. If these biotechnologies and their products can be validated with regard to such issues, then they have considerable potential to contribute to occupational cancer research.

Although workplace risks have various distinguishing characteristics, ultimately it may make more sense to consider work and non-work-related risks holistically. Workplace risks are caused primarily by exposure to workplace hazards; however, these risks are mitigated by various personal and lifestyle factors, such as environmental tobacco smoke. The role of workplace environment, including sedentary occupations and lack of access to healthy foods, has not been examined extensively. Workplace physical activity has consistently been found to be protective against colon cancer; conversely, sedentary occupations increase risk if not balanced with physical activity outside work.

Occupational factors appear to contribute to disparities in morbidity and mortality by demographic and socioeconomic factors. In addition to protecting the workforce from exposure to specific occupational hazards, there is a need to understand how the workplace environment interacts with other factors to produce the enormous gaps in health and mortality in the U.S., and how the workplace can play a role in closing those gaps. A holistic view of a disease, such as cancer, requires an understanding of all risk factors that could lead to it.

Control Technology and
Personal Protective Equipment

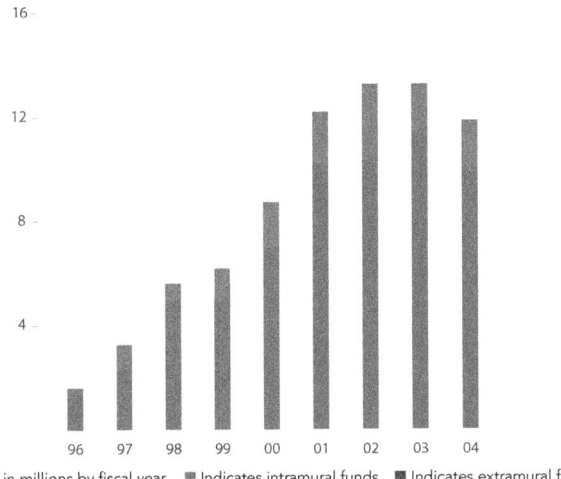

Funding in millions by fiscal year.   ■ Indicates intramural funds   ■ Indicates extramural funds

Thanks to a new low-cost ventilation exhaust system, workers near asphalt paving machines have greater protection from asphalt fumes. Substituting plastic for glass bottles is preventing low back disorders among workers who handle and transport beverages. Improved respirator filters provide greater protection in workplaces ranging from health care facilities to metal fabrication shops. As these examples illustrate, research in control technology and personal protective equipment can have widespread, direct impact on workers' safety and health.

A DECADE OF PROGRESS

Award-winning partnerships have been the hallmark of the NORA Control Technology and Personal Protective Equipment (CTPPE) team. The team's diverse membership regularly collaborated with groups like the American Industrial Hygiene Association (AIHA), the American Society of Safety Engineers (ASSE), and members of the health care community. Together, they determined basic and applied research needs for identifying, developing, and evaluating practical and effective control strategies. NORA engineering research in the asphalt paving industry was nationally recognized as a top 10 finalist for the Ford Foundation and Harvard University Innovations in American Government Awards in 1998 and was the first recipient of the NORA Partnering Award in 1999.

The team first gathered a broad consensus of research needs at a unique 1998 workshop, titled "The Control of Workplace Hazards for the 21st Century: Setting the Research Agenda." Co-sponsored with AIHA and ASSE, the conference assembled more than 250 researchers, manufacturers, and users of engineering controls and personal protective equipment. Together, they defined a future research agenda, which identified six priority areas: (1) biological and chemical protective clothing, (2) engineering controls, (3) noise, (4) non-ionizing radiation, (5) respirators, and (6) traumatic injuries. Important concepts of the framework include:

> Holistic or broad spectrum approach. Control technology should be used to reduce all workplace hazards, not just single hazards, such as air contaminants.

> Importance of primary prevention. Primary prevention seeks to eliminate the possibility of injury or disease rather than reducing the probability.

> Practical and cost-effective control technologies and interventions. Control technologies and interventions will not be used unless they are practical and cost-effective.

> Field-testing control technologies and interventions to evaluate their efficacy and compatibility to the industry and the workers.
>
> Additional training of undergraduate and graduate engineering, business, and architecture students in the primary prevention and control of environmental and occupational hazards.
>
> Communication of the results of control technology research to safety and health professionals, executives, supervisors, and workers.

After the creation of this common research framework, additional team partnerships emerged, resulting in important impacts. Government, industry, and labor partnerships reduced worker exposure to asphalt fumes and sponsored numerous symposia on work-related hearing impairment. The team also collaborated with the NORA Reproductive Health Research Team to create the NIOSH Hazardous Drug Safe Handling Working Group. This workgroup, formed from a partnership of more than 60 diverse members, produced an influential NIOSH Alert, titled *Preventing Occupational Exposures to Antineoplastic and Other Hazardous Drugs in Health Care Settings.* The workgroup recommended new guidelines to protect the nearly 5.5 million health care workers who are exposed to drugs known or suspected to cause health effects ranging from skin rashes to reproductive diseases, and possibly cancer.

Since the Alert was released in September 2004, six of seven manufacturers of equipment for pharmacy compounding have begun marketing engineering controls specific to the guidance introduced in the document. NIOSH-conducted studies were instrumental in forming an American Society for Testing and Materials standard for glove testing, which the Food and Drug Administration will use when approving gloves for use with chemotherapy drugs. The Alert also received the inaugural 2005 NIOSH Bullard-Sherwood Research-to-Practice (r2p) Award.

FUTURE DIRECTIONS

New control measures must be developed to address current and emerging workplace hazards, such as nanotechnology. Existing controls, even if effective, may lack acceptance or be perceived as cost-prohibitive. For jobs in which personal protective equipment is the only available option to ensure worker safety, controls must be developed that are effective and practical. The control also must not introduce a hazard greater than the one it is intended to prevent, a concern raised by health care workers who developed significant allergic responses to wearing latex gloves. Personal protective equipment also must be designed and made available to properly fit the growing numbers of female and minority workers.

Based on the six areas identified during the conference on "The Control of Workplace Hazards for the 21st Century: Setting the Research Agenda," the following future research recommendations are proposed:

Biological and Chemical Protective Clothing

Investigate dynamics of biological and chemical (B&C) dermal exposures.

Improve laboratory and field-testing methods.

Improve B&C protective clothing relative to human factors and ergonomics.

Promote better education and training of users regarding the selection, use, and limitations of B&C protective clothing.

Develop B&C protective clothing decontamination and disposal methods.

Engineering Controls

Review process-specific engineering controls for a variety of common hazards.

Apply process review techniques for safety management to control worker exposures.

- Study industrial hygiene implications of air-cleaner performance, reliability, and economics.
- Encourage use of computational fluid dynamics for design of local exhaust systems.
- Develop quantitative models relating cost of process changes to exposure reduction.
- Measure economic impact of pollution and hazard prevention, with particular emphasis on substitution for less hazardous chemicals.
- Integrate occupational safety and health principles into college engineering curricula.
- Develop better methods for information dissemination and training.
- Develop a clearinghouse for completed engineering control research to facilitate information exchange among the many groups affected.
- Conduct a needs assessment and identify research gaps for control banding as a control strategy in small businesses and developing countries.

Noise

- Assess the distribution of exposures and the current state of noise exposure monitoring through surveillance, with particular emphasis on impact noise.
- Develop new technologies to provide quieter machines, low-cost noise dosimeters, research on the redistribution of hearing conservation responsibility to the worker, and Material Safety Data Sheets (MSDS) for reference in the workplace.
- Research hearing protective devices (HPDs), the effect of HPDs on the workplace, new technologies integrating communication systems, and other technologies where HPDs are needed.
- Increase public awareness of hearing health as a wellness issue.

Non-Ionizing Radiation (NIR)

Improve instrumentation and techniques to address measurement and control of exposures in the near-field.

Improve engineering controls, personal protective equipment and monitoring instruments for dealing with NIR exposures in the workplace.

Encourage participation of both industrial hygienists and management to address NIR workplace hazards effectively.

Improve worker and safety professionals' awareness of NIR issues through training.

Respirators

Conduct an assessment of respirators used by industry and provide smaller businesses, especially, with recommendations for respirator use and training for the specific industry.

Standardize terms, recommendations, and protection factors that trigger actions in a respirator program.

Catalog end-of-service-life indicators for cartridges and filters that do not last indefinitely.

Research fit-testing criteria to learn the optimal duration and frequency of testing, how fit-testing results compare to workplace performance, and the cost-benefit of fit testing.

Traumatic Injuries

Transfer safety and technology research and practices to small businesses.

Build in product safety (safety by design) to protect workers.

Research the effect of psychosocial factors on occupational injury.

> Research traumatic injury controls for special groups (e.g., health care, farming, maritime, retail store, general office, and transportation workers).
>
> Research the effect of multi-risk factors (work task/environment, workforce/worker, and organization/management) on traumatic injury controls.
>
> Research material-handling methods/techniques and traumatic injury controls.

The team's research agenda identifies opportunities for research, partnership, intervention, and dissemination. These opportunities can be pursued through the industry- and cross-sector approach of NORA. The next decade of NORA provides the opportunity to expand and strengthen the existing partnerships among stakeholders to build on the accomplishments of the NORA CTPPE Team.

Exposure Assessment Methods

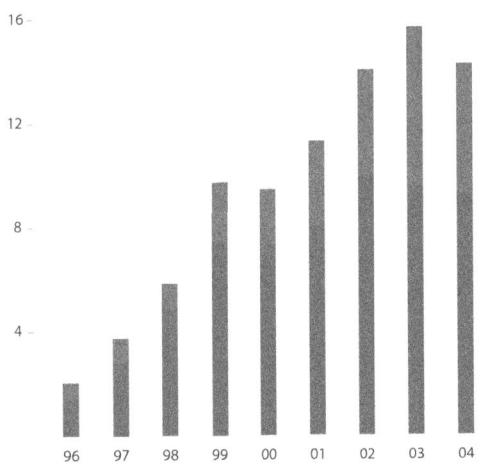

Funding in millions by fiscal year. ■ Indicates intramural funds ■ Indicates extramural funds

Accurate exposure data are fundamental to occupational safety and health research. Without effective ways to identify and measure harmful exposures, inaccurate research conclusions could occur. Diseases, for example, could be attributed to the wrong exposures, or important health effects could be overlooked. Exposure assessment research is a multidisciplinary field designed to improve scientific accuracy, define exposure response relationships, and reduce harmful workplace exposures.

The scope of occupational exposure assessment had broadened considerably in the years immediately preceding NORA. Rapid technological changes were occurring, and increased research attention was being placed on non-industrial work settings. Major gaps existed in the then current research methods, gaps that would eventually drive the development of research methods for exposure assessment. These gaps included a lack of sufficiently precise methods for exposure assessment to support accurate epidemiologic studies in complex work environments; practical measurement techniques that could be applied at reasonable cost; and validated methods for measuring relevant exposure and total dose data directly from biological samples obtained by relatively non-invasive techniques.

As NORA began, stakeholders believed that advancements in exposure assessment could help identify at-risk workers, develop the most cost-effective control and intervention strategies, illuminate exposure-response relationships, and improve baseline data for standard setting and risk assessment.

A DECADE OF PROGRESS

The NORA Exposure Assessment Methods Team began its work attempting to define exposure assessment. This exercise resulted in a holistic definition of exposure assessment that would be meaningful to all occupational health and safety disciplines and that would encourage the collection of more universally practical exposure assessment data. This effort resulted in a peer-reviewed manuscript, entitled *A Proposal to Standardize the Definition of Exposure Assessment*.

A primary goal of the team was to identify research gaps in exposure assessment. This effort led to the 2002 white paper *Exposure Assessment Methods—Research Needs and Priorities*. Within the white paper, the team proposed definitions and prioritized recommendations for research related to the field of exposure assessment. Proposed research falls into four areas: (1) study design, (2) monitoring methods, (3) applied toxicology, and (4) education and communication. The purpose of this document is to stimulate discussion and new research in those areas.

The recommendations in the white paper sparked a third major team effort and publication. The team sponsored and organized an international workshop on the development and use of biomarkers from toxicologic, risk management, ethical and legal, and regulatory perspectives. This workshop resulted in a proceedings publication in the *International Journal of Occupational Environmental Health* and sparked an additional manuscript, entitled *Biomarkers in Occupational Health Practice*. This paper discussed established biomarkers and their applications, new biomarker technologies and their potential roles, and gaps and barriers that prevent widespread use of biomarkers. In addition to the above-mentioned products, the team sponsored three other meetings and workshops and has five other publications.

FUTURE DIRECTIONS

New technologies, such as nanotechnology, genetic engineering, semiconductors, and electronics, have advanced beyond our knowledge or experience to anticipate, recognize, or evaluate exposure. Businesses are creating products faster than the health and safety community can develop or conduct appropriate methodologies for hazard characterization and exposure determination. This problem is compounded by the increase of small- to medium-size companies that have limited financial resources and professional expertise available to properly address occupational exposures. Proposed exposure assessment methodologies like control banding have limited validation and require toxicity information that may not be readily available.

In consideration of the above direction, several new priority areas should be added or at least existing priority areas should be expanded:

> Research is needed to advance and streamline the hazard and exposure assessment processes associated with emerging technologies.

> New methodologies will be required to better anticipate and recognize problem areas and to produce exposure assessments that address risk assessment, management, and communication issues.

> Control-banding methods should be enhanced, and additional modeling methods should be developed and validated to address other needs such as exposure classification, exposure ranking,

data interpretation, expert systems, and complex exposure scenarios such as mixtures and non-ambient conditions.

Methodologies between disciplines that rely on exposure assessment should be integrated.

Available information on the relationships between exposure and health outcomes is being gathered and published as a practical guide for epidemiological studies to improve the consistency and quality of exposure assessment, as well as to identify gaps in the knowledge base and opportunities for improvement of exposure assessment.

The National Occupational Exposure Survey should be updated and expanded.

The necessary contents of a national occupational exposure database are being developed.

A data interpretation and analysis guide should be developed. The AIHA has published a document, entitled A Strategy for Assessing and Managing Occupational Exposure, which advances the work on determining compliance with an occupational exposure limit. Further work is needed to determine sampling strategies so that the decision process can be based on modeling efforts.

Refinement of injury risk assessment, such as developing accurate exposure durations, will identify at-risk workers. It can also result in cost-effective control and intervention strategies, exposure-response relationships, and more accurate risk assessments.

Control banding, a qualitative predictive model, should be considered as a new approach to exposure assessment.

Methodology is needed to measure non-traditional exposures such as work organization and other stressors.

Methodology and reporting of results should be standardized to make it easier for all to understand and use exposure assessment data.

A biomarker database should be developed. This database would list the biomarker and the stage of validation.

A national biomonitoring manual should be developed from which standardized methods could be drawn.

Guidelines should be developed that could be used by occupational safety and health practitioners to use biomarker and guidelines to interpret and communicate biomarker results.

Health Services Research

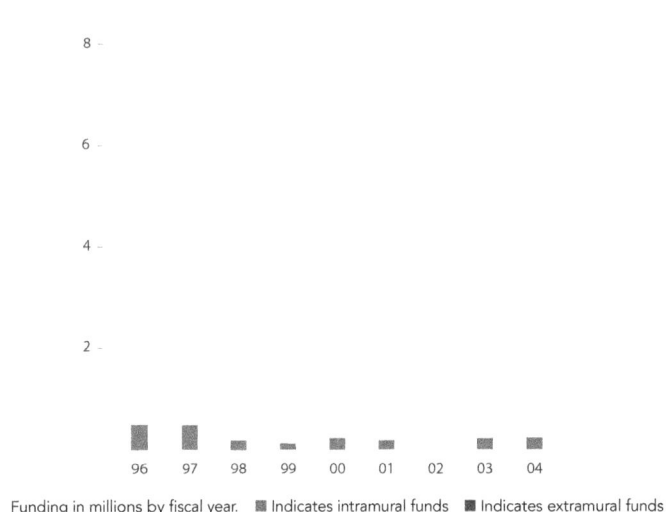

Funding in millions by fiscal year. ■ Indicates intramural funds ■ Indicates extramural funds

Health services research (HSR) explores the relationship between the type and quality of health services and the health outcomes of individuals and communities. HSR examines diverse areas such as health care management, economics, the type and delivery of care, and the effectiveness of health care policies, practices, or prevention strategies. While HSR had grown in scope and significance before NORA, little attention was paid to the association between health services and work-related injuries and illnesses.

The NORA HSR Team sought to promote better understanding of the contribution of occupational health services to the recognition, treatment, and prevention of work-related diseases and injuries. Diverse approaches were needed to address concerns about the quality of care, access to care, and the availability of well-trained health professionals and researchers. In addition, the team sought to understand the relationship between health service interventions and health, social, and economic outcomes.

A DECADE OF PROGRESS

Occupational HSR has developed since the start of NORA, beginning with an increase in training programs. In 1999, NIOSH announced a targeted training support program to address the dearth of trained investigators focused on work-related issues. Programs funded under the training grants were expected to train researchers, promote collaborative research, and build bridges between institutions that train researchers and the organizations involved in managing occupational health services. Four NIOSH Educational and Research Centers (ERCs) successfully competed to add this new component to their programs: Harvard University, University of Minnesota, University of Washington, and University of North Carolina (UNC). All have made progress toward the original goals of the program.

The availability of research funding attracted experienced occupational safety and health investigators as well as a limited number of general HSR investigators to the HSR field. NIOSH funded a range of investigations including studies to:

> Improve understanding of regional variations in treatment and outcomes of spinal cord injured workers;
>
> Expand use of workers compensation data for occupational surveillance;
>
> Investigate how different approaches to the diagnosis and treatment of ergonomic-related injuries affect outcomes;
>
> Develop evidence-based guidelines for occupational health practice; and
>
> Identify inequalities in access to occupational health services.

An important team accomplishment of the NORA HSR Team was a 1999 conference, co-sponsored with the NORA Social and Economic Consequences (SEC) team, the Robert Wood Johnson Foundation (RWJF), the Institute for Work and Health of Canada, and the Agency for Healthcare Research and Quality.

This conference gathered leading and emerging researchers to review the current state of knowledge and to discuss future research opportunities and priorities. Four of the HSR papers commissioned for the meeting were published in the *American Journal of Industrial Medicine* in 2001, along with most of the SEC background papers. These papers emphasize a number of key points raised during the conference, including the:

> Need for research to minimize disability and maximize injured workers' quality of life and labor market participation;
>
> Need for systematic collection and standardization of relevant data;
>
> Importance of clarifying what constitutes high-quality medical care for injured workers;
>
> Importance of examining a wide range of social, economic, and functional outcomes in addition to direct costs;
>
> Need to consider occupational health services in the broader context of general health services; and
>
> Importance of training investigators able to work successfully across a range of disciplines.

This conference also highlighted the need for an integrated view of health and safety risks. Traditionally, NIOSH had focused exclusively on risk from work. Occupational health services take a broader view by examining the relationships between workplace exposures, policies, and other health hazards, such as tobacco. The physical consequences of exposure to workplace hazards, such as asbestos and tobacco, have been well known for decades. Less attention had been paid to environmental tobacco smoke as a workplace health hazard and to the need for health services to assist cessation and workplace policies that promote prevention. Recent work has drawn attention to the unequal distribution of tobacco use among occupational categories and the importance of interventions, both in the workplace and out of work, that reflect that understanding.

The NIOSH Scientific Workshop titled "Work, Smoking, and Health," held June 2000, gathered a diverse group from academia, labor, industry, and governmental and non-governmental organizations to address questions about the impact of active or passive smoking, combined with occupational hazards, on the health of workers. The workshop also addressed the reasons for the uneven occupational distribution of tobacco use and the most effective policies and programs and priorities for future work. Workshop proceedings, background papers, and a summary of research recommendations were published as a NIOSH document in 2002.

The success of these meetings illustrates the importance of partnerships in advancing HSR. NIOSH and RWJF held a number of joint technical meetings and educational programs to promote exchange of information and research data. This collaboration grew from RWJF's Workers' Compensation Health Initiative, established in 1995 to develop and evaluate innovative approaches for the delivery of care to people injured at work. NIOSH and RWJF provided funding to researchers to attend these meetings and report on work in progress. This series of technical meetings played a critical role in advancing the field.

A final outgrowth of the Health Services Research NORA priority area has been the development of a new NIOSH and national program, titled the WorkLife Initiative. This initiative seeks to integrate efforts in health promotion and occupational safety and health in order to develop mutually supportive strategies for research and practice to improve worker health, safety, and well-being. In 2004, NIOSH and more than 20 co-sponsors organized a groundbreaking national symposium to explore integrating approaches to reduce occupational and personal health risks, the economic context of these programs, and the current practices that best address the combination of occupational workforce health protection and individual health promotion.

FUTURE DIRECTIONS

Once virtually absent from the NIOSH intramural and extramural programs, the WorkLife Initiative is expected to become a major theme of NIOSH's future work and supports the current NIOSH r2p focus.

Despite the many successes in this priority area, the need to develop cross-cutting research consistent with many topics explored in the earlier HSR/SEC recommendations still exists. Research is needed to address concerns about access to care for occupational injuries and illnesses, quality of care, availability of health professionals, and patterns of health care cost and utilization.

Intervention Effectiveness Research

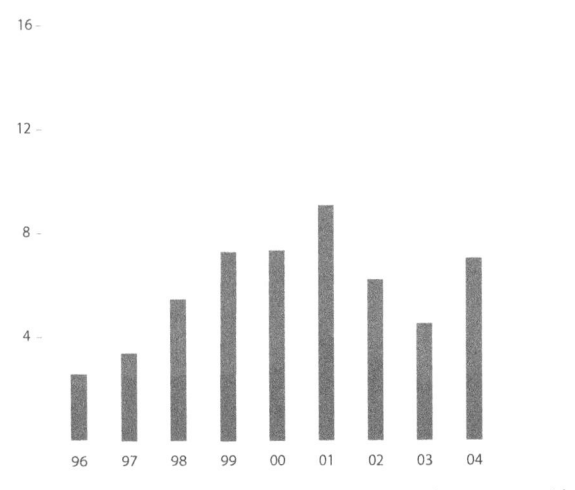

Funding in millions by fiscal year. ■ Indicates intramural funds ■ Indicates extramural funds

Intervention effectiveness research (IER) examines the connection between a precisely defined intervention and the results of using it in the workplace. An intervention may include preventive technologies, hazard exposure guidelines or standards, reorganization or restructuring of work processes, and administrative changes such as job rotation or training. Knowledge about what prevents workplace illness and injury is useful for economic, standard-setting, and public health purposes. Measuring the results of planned actions is a recognized part of good management. Though critically important for research and practice, IER was an underutilized tool when NORA began a decade ago. The NORA IER Team adopted an ambitious agenda of outreach and research to promote this priority area.

A DECADE OF PROGRESS

The team members published a conceptual paper on the IER process. The team believes that the complete this process includes activities related to development of the intervention, its implementation, and the measurement of its effectiveness. Within each of these stages, a five-phase subprocess occurs, which consists of gathering background information; developing partnerships; choosing methods and designs; completing the development, implementation, or evaluation; and reporting and disseminating findings.

The team collaborated with researchers at the Institute for Work & Health (IWH) in Ontario, Canada, to develop a manual of recommended IER methods in occupational safety and health, titled *Guide to Evaluating the Effectiveness of Strategies for Preventing Work Injuries*. This resource guide for researchers was completed and published jointly by NIOSH and the IWH in 2001. The team also prepared a booklet on IER for non-researchers, titled *Does It Really Work?*, which NIOSH published in 2004. Both publications have enjoyed a wide readership.

The team also cooperated with a number of organizations to encourage IER and promote written guidelines. These partnerships crossed a number of industries and occupational groups, including state health departments, safety and health professionals in agriculture, safety professionals in the National Safety Council (NSC), and the Maine Occupational Research Agenda. Most often, the partnership involved presenting workshops or seminars on IER. The team developed targeted curricula that addressed the learning needs of each group. The curricula featured active, problem-solving activities that required the application of study design principles to occupational safety and health problems that were relevant to each particular audience. With team support, the NSC developed and presented a professional development seminar in IER. A pilot version was presented and evaluated at the NSC's annual conference in 2002. Since then, this day-long seminar has been refined and presented at each subsequent NSC annual conference.

NIOSH, the IER team, and its partners also sponsored an international contest to recognize outstanding evaluations of interventions designed to prevent work-related injuries. Contest rules allowed for evaluations of any intervention ranging from a simple tool or equipment change at a single worksite to programs of training to protective equipment use. The intervention could also be of one or more policies applied in a corporation, an industry, a State, or a nationality (e.g., an OSHA standard). The contest winner and two honorable mention submissions were awarded plaques and provided travel assistance to present their work at a special session of the National Occupational Injury Research Symposium (NOIRS) on October 28, 2003. In addition to the presentations by contest winners, the special session also discussed the challenges of conducting IER work and presented solutions to these challenges.

The team encouraged collaboration with researchers in other NORA priority areas. Team members attended meetings of the Hearing Loss, Musculoskeletal Disorders, Special Populations at Risk, and Reproductive Health Research teams. Team members supported a workshop conducted by the Musculoskeletal Disorders Team on ergonomic IER in 2004. A meeting with the Reproductive Issues Team resulted in an effort to evaluate the effectiveness of an information product, a NIOSH Alert on exposure to hazardous drugs in health care settings. The team also is advising a group of NIOSH researchers about evaluating interventions that target special populations.

Finally, the team is working with NIOSH to develop a database of interventions and an intervention dissemination Website for small businesses. Such resources are currently scattered. Descriptions of interventions that do exist are often not accessible to wide audiences. Effectiveness data are often not presented or non-existent, leaving readers to wonder whether an intervention will really work or not. The Center to Protect Workers' Rights (CPWR) and NIOSH are partnering on the development of the

Website structure and format in order to harmonize the presentation and offer standardized evaluations of the solutions. Users will be able to search the database for hazard information, training materials, and other potential interventions. The first focus will be on construction industry interventions. Future efforts will include agriculture and mining. The IER team will evaluate the quality of the effectiveness research evidence available for each intervention proposed for inclusion in the database.

Clearly, there have been many developments, both theoretical and practical, in IER since the beginning of NORA. Other indicators also point to change. During the NORA years, NIOSH announced at least four requests for grant applications for IER in construction safety and health (1997), agricultural safety and health (1997), occupational safety and health (2000), and violence prevention research (2002).

There was a substantial increase in active federal grants in IER after 1996. Although there are many possible explanations for this result, these data are not inconsistent with the possibility that NORA (and its IER team) stimulated more research. Goldenhar and Schulte (1994) published a review of the occupational safety and health (OSH) intervention effectiveness literature. They found 34 studies that occurred during the 6-year period from 1988 to 1993. In 2003, the IER team found more than 100 OSH intervention effectiveness studies published during the 6-year period from 1998 to 2003.

FUTURE DIRECTIONS

A strength of the NORA team concept was assembling researchers who are doing similar work, but come from different perspectives. This collaboration increased the team's awareness of the close relationship between IER and important safety and health changes in the workplace. The group felt the urgency and importance of increasing and improving IER practice. Improving the quality of IER through methods standardization had to be secondary to improving overall practice and moving IER results into the workplace. The following recommendations carry on that practical perspective:

- Continue to improve and disseminate practice guidelines.
- Strengthen the involvement of stakeholders in designing and conducting IER.
- Encourage the conduct of policy intervention studies.
- Study the process of transfer of IER results from evaluator to user.
- Continue to develop low-tech IER methods.
- Encourage IER in all NORA research areas.
- Ensure that IER methods are being taught in the classrooms of the OSH professions.

Risk Assessment Methods

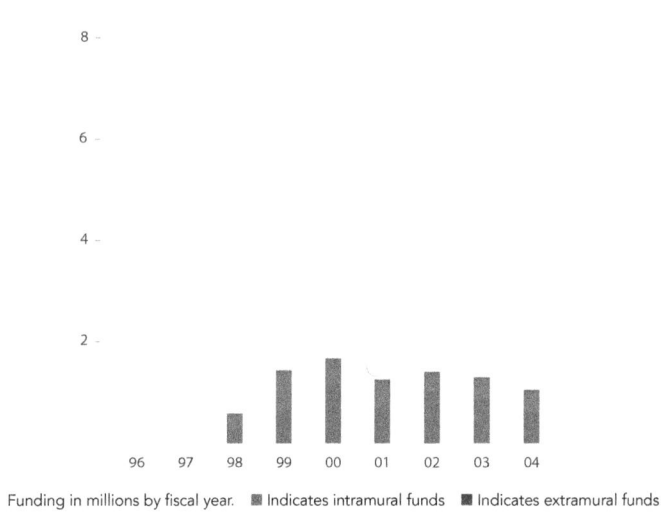

Funding in millions by fiscal year. ■ Indicates intramural funds ■ Indicates extramural funds

Risk assessment is the process that identifies hazard, exposure, and dose-response information to determine whether an exposed worker population is at greater-than-expected risk of disease or injury. The magnitude and nature of the increased risk can then be investigated, using either qualitative or quantitative approaches. Qualitative risk assessments are generally descriptive and indicate that disease or injury is likely or unlikely under specified conditions of exposure. Quantitative risk assessments provide a numerical estimation of risk based on mathematical modeling, such as 1 person per 1,000 would be expected to develop a disease or injury given specific exposure conditions.

Quantitative methods require both detailed data on relevant exposures and a mathematical model to describe the exposure-response relationship. Animal models and epidemiology studies can provide such data, but limitations exist. For example, experimental animal and molecular biologic data provide detailed information on the exposure-response relationships, but workers may have much lower and more variable exposures than susceptible animal species tested at constant high doses. Risk assessments based on population-based studies may have real-world relevance to workers, but they generally suffer from a number of limitations, including potential confounding by risk factors or other exposures; variability in workplace exposures; individual variability in health response; and detection of statistically significant changes in adverse health outcomes.

Quantitative risk assessments may often be preferred whenever data, modeling techniques, and biological understanding are adequate to support their development. A call for more formal quantitative methods for occupational safety and health regulation began when the U.S. Supreme Court ruled in the "benzene decision" (Industrial Union Department v. American Petroleum Institute, 448 U.S. 607 [1980]). This decision stated that OSHA could not issue a standard without demonstrating a significant risk of material health impairment. The ruling encouraged, but did not demand, the use of numerical criteria to determine whether a risk is "significant." As a result of that Supreme Court ruling, risk assessment became standard practice in OSHA rulemaking for health standards.

Yet, when NORA began in 1996, considerable controversy surrounded many of the risk assessment methods for cancer and non-cancer effects, and methods for assessing safety risks were perceived as even less developed. NORA stakeholders recognized a wide range of scientific disciplines were essential for providing more reliable methods for estimating work-related risks. The NORA Risk Assessment Methods (RAM) team was formed to identify opportunities for new approaches to modeling and risk communication.

A DECADE OF PROGRESS

During the last 10 years, risk assessments have become increasingly sophisticated, sparking even greater support for formal risk assessments in establishing national priorities and in justifying regulatory actions taken by federal agencies. Given the important emphasis placed on risk assessment, the NORA RAM Team sought to identify research areas needed to improve methodologies. The team identified five research areas, including:

- Sources of human variability in susceptibility among workers exposed to toxic substances;
- Evaluation of toxicological risk assessment models using epidemiological data;
- Use of biomarkers of disease to develop risk assessment models for predicting the risk of chronic diseases;
- The impact of errors in exposure assessments on epidemiologic risk assessments; and
- The influence of pattern of exposures in determining occupational health risks.

NIOSH supported these research priorities by funding evaluation research for toxicological and epidemiological estimates of the risk for carcinogenic hazards. In addition, NIOSH, EPA, and NCI funded a 1999 RFA for research focusing on the development of cancer risk assessment methods and practices.

Perhaps the most important accomplishment of the RAM team was to sponsor a workshop on "Future Research to Improve Risk Assessment Methods" in August 2000. The primary purpose of this workshop was to develop a national research agenda for risk assessment methods. Working groups prepared research recommendations to improve methods for conducting quantitative risk assessments in the areas of toxicology, epidemiology, and modeling. *The Journal of Human and Ecological Risk Assessment* published the conference proceedings in 2002.

The toxicology workgroup recommended research to do the following:

> Increase understanding of inter- and intra-individual variability in susceptibility during all life stages.
>
> Account for factors that affect cross-species extrapolation.
>
> Adjust for dose rate effects.
>
> Define toxicological responses at low doses.
>
> Make better use of continuous as well as quantal data from toxicological responses.
>
> Develop better response data from short-term exposures.
>
> Address exposures to chemical mixtures and by multiple exposure routes.
>
> Refine uncertainty factors with reliable experimental data.

The epidemiology group concluded methodological research is needed in the following areas:

> Aspects of epidemiologic study designs that affect dose-response estimation;
>
> Alternative methods for estimating dose in human studies; and
>
> Refined methods for dose-response modeling for epidemiologic data.

The modeling group recommended research to do the following:

> Characterize inter-individual and inter-species variability in susceptibility.
>
> Develop improved models for injury risk assessment, including development and characterization of risk estimates and exposure metrics.
>
> Adapt and modify existing standard procedures (e.g., to derive reference doses).

- Develop acceptability criteria for mechanistic hypotheses and data.
- Develop models for multiple endpoint data, particularly for different endpoints on different scales.
- Develop new mechanistic models of carcinogenesis.
- Improve methods for combining data of different types in risk analyses.
- Explore evidence and models for complex dose-response relationships in the context of homeostasis (including models of hormesis).

FUTURE DIRECTIONS

Although research is ongoing in many priority areas, none have been definitively addressed. Researchers, for example, can better evaluate the concordance between toxicological and epidemiological estimates of carcinogenic risks, but challenges remain. Exposures measured in most epidemiological studies are relatively less precise than exposures measured under the controlled experimental conditions possible in toxicological studies. This lack of precision makes it difficult to determine whether toxicologically based risk estimates accurately predict the risk in humans. Toxicological studies, however, might not exist for substances with good quantitative epidemiology.

As with concordance issues, scientists have made distinct progress understanding inter- and intra-individual variations in chemical susceptibility, but many questions remain. These questions may be partially addressed by studies of inter-individual variations in xenobiotic metabolism and by future knowledge gained in the genomics revolution. The field of quantitative risk assessment is entering a new era of understanding the molecular basis for chemically induced disease. Genomic and proteomic data may provide important new insights, such as:

- Illuminating the genetic basis for individual variations in susceptibility;

- Identifying factors affecting cross-species extrapolation;

- Refining uncertainty factors;

- Contributing to new epidemiologic study designs that improve dose-response estimation;

- Developing improved biomarkers;

- Providing insight into the causative linkages between exposure and response;

- Improving our understanding of toxicological responses at low doses;

- Addressing exposures to chemical mixtures and by multiple exposure routes;

- Leading to alternative methods for estimating dose in human studies; and

- Modifying current mechanistically based models of carcinogenesis.

Despite clear needs and new opportunities, it is difficult to predict the future of risk assessment methods. Improvements are typically sparked by the need to address practical problems. Applied projects that address research gaps are needed for risk assessment and regulatory actions. In addition, methodological projects that address issues in both epidemiology and toxicology, particularly related to the use of genomic and proteomic data, should be supported. Improvements in the research base and the quality of epidemiological and toxicological methods would clearly be the best possible investment for enhancing our ability to accurately characterize human risks.

Social and Economic Consequences of Workplace Illness and Injury

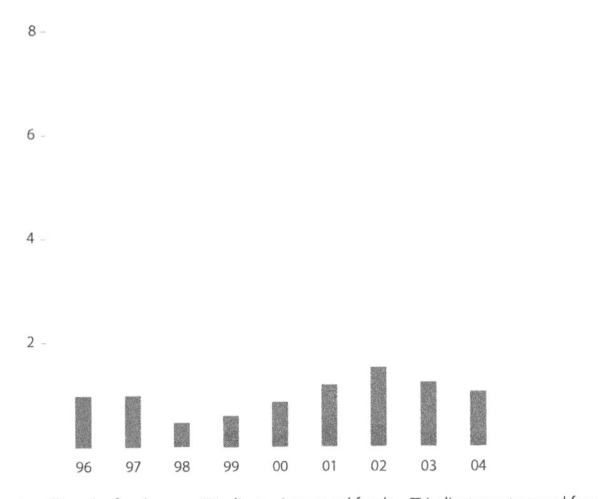

Funding in millions by fiscal year. ■ Indicates intramural funds ■ Indicates extramural funds

When workplace illnesses and injuries occur, workers' compensation claims are not the only result. Workers, their families, their employers, and society suffer from hidden costs such as decreased quality of life, increased time and work demands on worker families, disruption of work, reduced productivity, and costs to the health care system and programs for the disabled. In 1996, NIOSH and its partners identified social and economic consequences (SEC) as a priority NORA research area. When the SEC team was formed, there was limited SEC research in the U.S., with the costs of workers' compensation insurance and claims being the largest single area of focus. The team sought to foster research that would identify and understand the complete costs of workplace illnesses and injuries.

A DECADE OF PROGRESS

In 1999, the team worked to define the direction of SEC research in a conference co-sponsored by the NORA Health Services Research Team, the Robert Wood Johnson Foundation, the Institute for Work and Health of Canada, and the Agency for Health Care Research and Quality. In 2001, the *American Journal of Industrial Medicine* published a set of papers from this conference, in what was, at the time, the most direct and in-depth treatment of the subject matter in an occupational safety and health journal.

Several themes emerged that clarify the state of knowledge, research gaps, and needs, and are summarized below:

> The economic burden of injuries on workers is very great, and more research is needed to determine the non-monetary costs.
>
> Psychologists, sociologists, anthropologists, and others have begun to describe social consequences of work injuries, including depression, a reduced ability to perform social and family roles, and difficulty maintaining family relationships.
>
> Almost all analyses of employer costs have focused on worker's compensation claims, overlooking less visible costs, including hiring and training of replacements and the impact on productivity.
>
> Workers tend to return to work earlier when they return to the same employer, and return-to-work programs can improve long-term work outcomes in ways that future studies should aim to quantify.
>
> More research is needed on factors that discourage workers from filing workers' compensation claims, resulting in under-reporting of illnesses and injuries.
>
> Workers' compensation benefits often fall short of replacing lost earnings and do not cover all medical costs.

> There is much unrealized potential for using existing administrative and survey data to assess economic and social consequences. Additionally, new data need to be developed using better measures of causal factors and outcomes.

NORA has increased funding for SEC research to address these knowledge gaps, with substantial focus on both workers and employers. Worker-focused research explored lost earnings, changes in employment status, medical costs, impacts on the worker's family, and the overall value to workers of avoiding accidents. Employer-focused research examined worker compensation costs and prevention measure costs, whereas a few studies looked at costs associated with productivity losses. A smaller number of projects focused on societal costs, such as costs to public programs like Social Security Disability Insurance (SSDI) and Medicare.

The primary focus of approximately one third of the economics-related projects was not economic questions, but identifying the impact of risk factors on illness and injury or measuring intervention effectiveness. This finding suggests that economic research can often be most effectively done in the context of more traditional occupational safety and health research. Additionally, a significant number of the extramural projects dealt with economic determinants of workplace injury and illness rather than their economic consequences, focusing on such things as availability of economic resources for prevention, competitive pressure on employers, and the economic situation of workers as a factor affecting time away from work after an injury. Future economic research should continue to examine these kinds of subjects.

The team also has sought to increase the visibility of SEC research while bringing together other stakeholders. For example, the team produced a partially annotated bibliography of the social consequences of workplace injury and illness (NIOSH, 1999), and a database of researchers of social and economic consequences (NIOSH, 2000), and sponsored production of a compendium of NIOSH-supported projects with economic components (NIOSH,

2005). One of the team's last major accomplishments was co-sponsoring, along with the NORA Intervention Effectiveness Research Team and other organizations, a conference organized by NIOSH and WHO, entitled *Economic Evaluation of Occupational Health and Safety Interventions at the Company Level.* This meeting was the first to bring together individuals who have independently developed methods for analyzing employers' investments in safety and health, and was attended by representatives from international governments, academics, large and small businesses, research and professional organizations, and NIOSH researchers.

NIOSH has a growing commitment to economics research. During the last decade of NORA, NIOSH hired four economists and two economist research fellows. It supported an internal Economics Interest Group that evolved in 2004 into the NIOSH Economics Forum, which supports the development and diffusion of economics research within the Institute and positions NIOSH to address future research needs.

FUTURE DIRECTIONS

The team has drafted a research agenda that evolved from the 1999 conference. Its major themes are outlined below:

> More research is needed on employer costs and consequences, especially for costs other than workers' compensation. Tools for employer evaluation of prevention measures also must be further developed.

> Methods of economic evaluation at all levels—worker, employer, and societal—must be developed and standardized.

> Sources of support for injured workers must be evaluated for their adequacy, equity, cost, and effectiveness, including workers' compensation, disability insurance, medical insurance, SSDI, and Temporary Assistance to Needy Families (TANF).

> Research is needed to document how workers and their families cope with a disabling work

injury, including how they alter their patterns of activity and work, and how their mental health and relationships are affected.

The long-term effects of injury and illness on workers and employers due to recurrence of health problems, change in jobs, or withdrawal from the workforce need to be studied.
The role of many factors in determining the ultimate success of workers in returning to work needs further study: the nature of the injury or illness, the work environment, the state of the labor market, and employer and workers' compensation policies.

Work injuries and illness reduce the nation's productive capacity and divert resources to health care. Full assessment of the economic burden of occupational injuries and illnesses requires estimation of their impact on gross domestic product (GDP), tax revenues, growth, competitiveness, and income distribution.

Research is needed on how the burdens of occupational injury and illness are shared by workers, family members, employers, the wider community, and government. This research promises to improve incentives for greater prevention.

Examination of the consequences of workplace illness and injury by specific condition, injury type, and hazard, and the degree to which they may be greater for certain vulnerable worker groups is needed in order to set research and prevention priorities.

Surveillance Research Methods

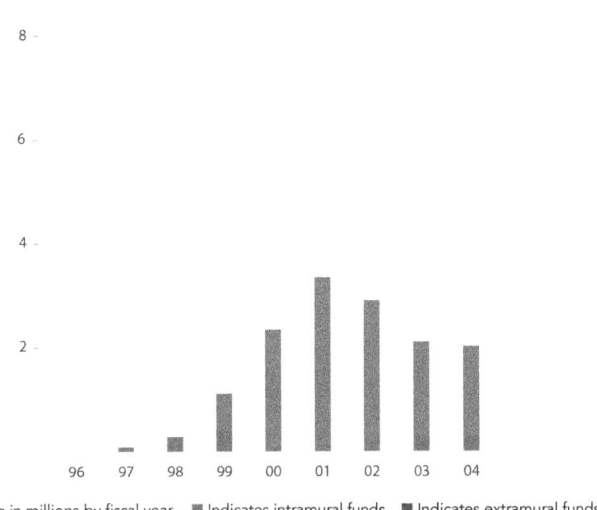

Funding in millions by fiscal year. ■ Indicates intramural funds ■ Indicates extramural funds

Surveillance is the collection, analysis, interpretation, and dissemination of data describing a health- related event. Surveillance is critical to effective occupational safety and health programs. It enables decision makers to identify the problem and the affected group of workers. Surveillance also describes the magnitude and severity of an issue and assesses progress made in reducing the burden of occupational injuries and illnesses. As a result, surveillance programs create added value by establishing baseline and trend data, assisting in priority setting, and providing information to guide research, interventions, control, or prevention.

Surveillance research methods are inherent to the development and continuing support of health surveillance activities. These activities include a range of epidemiologic and statistical methods, tools, approaches, and strategies integral to the development, maintenance, and enhancement of public and occupational health surveillance programs. These methods are developed or adapted to estimate, enumerate, and describe health-related events, worker characteristics, workplace environments, and the morbidity, mortality, and health status of workers in the U.S.

Surveillance and survey methods were well established through the mid-1990s for monitoring injuries, illnesses, and fatalities on a national level. Successful federal-state partnerships supported ongoing fatality evaluations, and case-based surveillance methods developed between 1987 and 1992 were being tested and evaluated in states for a variety of occupational concerns. Four challenges, however, inhibited continued state-of-the-art advancements:

> Problems of illness and injury under-recognition and under-reporting;
>
> Inadequate or non-existent denominator and economic data;
>
> Inadequate or non-existent data on priority or vulnerable populations; and
>
> A lack of methods and strategies to identify and track problems among small-scale businesses, independent contractors, and part-time and occasional laborers.

A DECADE OF PROGRESS

Progress in developing surveillance research methods is evident in a number of team agenda-setting activities.

The NORA Surveillance Research Methods Team was a major contributor of input for the development of the NIOSH Surveillance Strategic Plan, advocating (1) strengthening the case-based surveillance capacity of state health departments and other state agencies to conduct occupational surveillance, and (2) promoting effective occupational safety and health surveillance conducted by employers, unions, and other non-governmental organizations.

Team members contributed to *The Role of States in a Nationwide Comprehensive Surveillance System for Work-related Diseases, Injuries and Hazards* and *Occupational Health Indicators: A Guide for Tracking Occupational Health Conditions and Their Determinants* reports. The latter proposes 19 surveillance indicators to guide states regarding the minimum level of occupational health surveillance activity. The indicators represent a core set of data that would assist each state with the development of programs to prevent workplace injuries and illnesses. NIOSH issued a Program Announcement to provide states the opportunity to develop these surveillance indicators to profile the state's population health status.

In 2001, the team convened a workshop on *Best Practices in Workplace Surveillance.* This workshop was a first step in implementing the fourth goal of the NIOSH Surveillance Strategic Plan, to promote effective occupational safety and health surveillance as conducted by employers, unions, and other non-governmental organizations. A workshop CD-ROM is available at the NIOSH Website. Workshop recommendations included:

> Expanded surveillance research at national and state levels;
>
> Adapting or scaling (for small workplaces) of the surveillance tools and techniques of larger employers;
>
> Applied surveillance research, integrating the principles of surveillance, risk management, and prevention; and
>
> Improve the dissemination research data.

NIOSH co-sponsored a workshop at the 2003 *Truck Driver Occupational Health and Safety Conference.* This workshop stimulated continued collaboration between NIOSH and transportation experts, leading to NIOSH's funding the Occupational Motor Vehicle Safety and Health Research program.

The team co-sponsored the 2004 workshop on Data Resources for Occupational Health Surveillance and Research focusing on the availability of denominator data for specific industry or occupation sectors; population-based data or establishment-based data for specific industry or occupation subgroups; how these data are made available or accessed; and whether these surveys or programs are amenable to "special supplements" to advance emerging research and other method development needs (i.e., special populations surveillance and research).

The peer-reviewed literature reflects progress to disseminate surveillance data and develop new surveillance systems. Although the problems of illness and injury under-recognition and under-reporting remain, surveillance data increasingly are used to identify new research opportunities, evaluate emerging health issues, and evaluate under-reporting. Although inadequate or non-existent denominator data remain problems for occupational disease mortality surveillance, population-based illness and injury surveillance programs and surveys are adopting methods that overcome or avoid these limitations. Recent NIOSH reports demonstrate the adaptability of current surveillance programs to design and conduct surveys on emerging safety and health issues and priority populations.

FUTURE DIRECTIONS

Maintain a strong national surveillance program to establish priorities. Future surveillance should (1) maintain ongoing surveillance and disseminate surveillance data as guided by the NIOSH Surveillance Strategic Plan, and (2) respond to emerging occupational health and hazard issues.

Create new program initiatives and projects to develop and adapt methods for state and non-governmental partners. New surveillance programs and research methods are reflected in the NIOSH Surveillance Strategic Plan and NORA research priorities for cancer, emerging technologies, exposure assessment methods, musculoskeletal disorders, traumatic injury, reproductive outcomes, and workplace organization factors.

Link the results from state-level surveillance to intervention and prevention activities. This step could produce significant improvements in occupational safety and health. Recent evaluation and planning activities reinforce the importance of expanding and enhancing state-based occupational surveillance.

Advocate an expanded surveillance research program that focuses on smaller employment establishments in a private-sector surveillance research initiative. An estimated 7 million private-sector establishments employed 115 million workers in 2001. Establishments with 19 or fewer employees accounted for 85.7% of all workplaces, but only 24.1% of all employees. Establishments with 100 or more employees accounted for only 0.7% of all workplaces, but over 46.8% of all employees.

Establish Collaborating Surveillance Research Centers of Excellence to guide the development of surveillance to prevention practices including new Research and Development teams that harness the strengths of occupational health researchers, non-government organizations, insurance carriers, and public health agencies. Specific activities within the Centers should include: (1) providing technical assistance and consultation with respect to developing and evaluating occupational surveillance methods; (2) establishing outreach programs to identify specific methodological and research needs, evaluate occupational surveillance follow-up methodologies, and develop and evaluate innovative strategies for improving the quality and utility of surveillance data; and (3) expanding surveillance and surveillance research that focuses on smaller-scale employment establishments.

APPENDIX

NORA Team and Related Publications: 1996-2006

Asthma and Chronic Obstructive Pulmonary Disease
Balmes J, Becklake M, Blanc P, Henneberger P, Kreiss K, Mapp C, Milton D, Schwartz D, Toren K, Viegi G [2003]. American Thoracic Society statement: occupational contribution to the burden of airway disease. Am J Respir Crit Care Med 167(5):787-797.

Leigh JP, Romano PS, Schenker MB, Kreiss K [2002]. Costs of occupational asthma and COPD. Chest 121(1):264-272.

Cancer Research Methods
Toraason M, Albertini R, Bayard S, Bigbee W et al. [2004]. Applying new biotechnologies to the study of occupational cancer: a workshop summary. Environ Health Perspect 112(4):413-416.

Ward EM, Schulte PA, Bayard S, Blair A, Brandt-Rauf P et al. [2003]. Priorities for development of research methods in occupational cancer. Environ Health Perspect 111(1):1-12.

Control Technology and Personal Protective Equipment
NIOSH [2004]. NIOSH Alert: preventing occupational exposures to antineoplastic and other hazardous drugs in health care settings. Cincinnati, OH: U.S. Department of Health and Human Services, Centers for Disease Control and Prevention, National Institute for Occupational Safety and Health, DHHS (NIOSH) Publication No. 2004-165.

NIOSH [2001]. Best practices in hearing loss prevention. Cincinnati, OH: U.S. Department of Health and Human Services, Centers for Disease Control and Prevention, National Institute for Occupational Safety and Health, DHHS (NIOSH) Publication No. 2001-157.

NIOSH [1997]. Engineering control guidelines for hot mix asphalt pavers. Cincinnati, OH: U.S. Department of Health and Human Services, Centers for Disease Control and Prevention, National Institute for Occupational Safety and Health, DHHS (NIOSH) Publication No. 97-105.

Exposure Assessment Methods
Abell MT, Woebkenberg ML, Armstrong TW, Stenzel M [2001]. Research recommendations of the NORA Exposure Assessment Methods Team. Appl Occup Environ Hyg 16(2):331-333.

NIOSH [2002]. Exposure assessment methods: research needs and priorities. Cincinnati, OH: U.S. Department of Health and Human Services, Centers for Disease Control and Prevention, National Institute for Occupational Safety and Health, DHHS (NIOSH) Publication No. 2002-126.

Woebkenberg ML [2000]. Gas and vapor exposure assessment methods. Appl Occup Environ Hyg 15(1):97-9.

Woebkenberg ML [2000]. Partnership in research: exposure assessment methods. Chemical Health & Safety *January/February:24-26.*

Hearing Loss
Morata TJ [2003]. Chemical exposure as a risk factor for hearing loss. Occup Environ Med. 45(7):676-82.

NIOSH [2001]. Proceedings: best practices in hearing loss prevention. Cincinnati, OH: U.S. Department of Health and Human Services, Centers for Disease Control and Prevention, National Institute for Occupational Safety and Health, DHHS (NIOSH) Publication No. 2001-57.

Health Services Research
Deitchman S, Dembe A, Himmelstein J [2001]. Advent of occupational health services research. Am J Ind Med 40(3):291-294.

Mustard C, Hertzman C [2001]. Relationships between health services outcomes and social and economic outcomes in workplace injury and disease: data sources and methods. Am J Ind Med 40(3):335-343.

NIOSH [2002]. Work, smoking, and health: a NIOSH scientific workshop. Cincinnati, OH: U.S. Department of Health and Human Services, Centers for Disease Control and Prevention, National Institute for Occupational Safety and Health, DHHS (NIOSH) Publication No. 2002-148.

Pransky G, Benjamin K, Dembe AE [2001]. Performance and quality measurement in occupational health services: current status and agenda for further research. Am J Ind Med. 40(3):295-306.

Rudolph L, Deitchman S, Dervin K.[Rudolph 2001]. Integrating occupational health services and occupational prevention services. Am J Ind Med. 40(3):307-318.

Shannon S, Robson L, Sale J. [2001]. Creating safer and healthier workplaces: Role of organizational factors and job characteristics. Am J Ind Med. 40(3):319-334.

Indoor Air
Hood E [2005]. Investigating indoor air. Environ Health Perspect 113(3)

Mendell MJ, Fisk WJ, Kreiss K, Levin H, Arch B, Alexander D, Cain WS, Girman JR, Hines CJ, Jensen PA, Milton DK, Rexroat LP, Wallingford KM [2002]. Improving the health of workers in indoor environments: priority research needs for a national occupational research agenda. Am J Public Health 92(9):1430-1440.

NIOSH [2006]. Indoor work environments and health: a research agenda. Cincinnati, OH: U.S. Department of Health and Human Services, Centers for Disease Control and Prevention, National Institute for Occupational Safety and Health, DHHS (NIOSH) Publication No. 2006-120.

Weschler C, Wells R. [2004]. Guest Editorial. *Indoor Air* 14(6):373-375.

Intervention Effectiveness Research
NIOSH [2001]. Guide to evaluating the effectiveness of strategies for preventing work injuries: how to show whether a safety intervention really works. Cincinnati, OH: U.S. Department of Health and Human Services, Centers for Disease Control and Prevention, National Institute for Occupational Safety and Health, DHHS (NIOSH) Publication No. 2001-119.

NIOSH [2004]. Does it really work? How to evaluate safety and health changes in the workplace. Cincinnati, OH: U.S. Department of Health and Human Services, Centers for Disease Control and Prevention, National Institute for Occupational Safety and Health, DHHS (NIOSH) Publication No. 2004-135.

Mixed Expsosures
NIOSH [2001]. Mixed exposures research agenda: a report by the NORA Mixed Exposures Team. Cincinnati, OH: U.S. Department of Health and Human Services, Centers for Disease Control and Prevention, National Institute for Occupational Safety and Health, DHHS (NIOSH) Publication No. 2005-106.

Musculoskeletal Disorders
NIOSH [2001]. National Occupational Research Agenda for Musculoskeletal Disorders: research topics for the next decade; a report by the NORA Musculoskeletal Disorders Team. Cincinnati, OH: U.S. Department of Health and Human Services, Centers for Disease Control and Prevention, National Institute for Occupational Safety and Health, DHHS (NIOSH) Publication No. 2001-117.

Organization of Work
Lamberg L [2004]. Impact of long working hours explored. JAMA 292(1):25-6.

Landsbergis PA [2003]. The changing organization of work and the safety and health of working people: a commentary. J Occup Environ Med 45(1):61-72.

Murphy LR, Sauter SL [2004]. Work organization interventions: state of knowledge and future directions. Soz Praventivmed 49(2):79-86.

NIOSH [2002]. The changing organization of work and the safety and health of working people—knowledge gaps and research needs. Cincinnati, OH: U.S. Department of Health and Human Services, Centers for Disease Control and Prevention, National Institute for Occupational Safety and Health, DHHS (NIOSH) Publication No. 2002-116.

NIOSH [2004]. Overtime and extended work shifts: recent findings on illnesses, injuries and health behaviors. Cincinnati, OH: U.S. Department of Health and Human Services, Centers for Disease Control and Prevention, National Institute for Occupational Safety and Health, DHHS (NIOSH) Publication No. 2004-143.

Sauter SL, Murphy LR [2003]. Monitoring the changing organization of work: international practices and new developments in the United States. Soz Praventivmed 48(6):341-8; discussion, 349-60.

Stellman JM [2001]. Filling the knowledge gap. The Synergist 12(9):v-vi.

Reproductive Health Research (Fertility & Pregnancy Abnormalities)
Grajewski B, Coble J, Frazier L, McDiarmid M [2005]. Occupational exposures and reproductive health: 2003 Teratology Society Meeting Symposium summary. Birth Defects Res B Dev Reprod Toxicol 74(2):157-63.

Lawson CC, Grajewski B, Daston GP, Frazier LM, Lynch D, McDiarmid M, Murono E, Perreault SD, Robbins WA, Ryan MAK, Shelby M, Whelan EA. [2006]. Implementing a national occupational reproductive research agenda: Decade one and beyond. Environ Health Perspect; 114:435-441.

Lawson CC, Schnorr TM, Daston GP, Grajewski B, Marcus M, McDiarmid M, Murono E, Perreault SD, Shelby M, Schrader SM [2003]. An occupational reproductive research agenda for the third millennium. Environ Health Perspect 111(4):584-592.

Moorman WJ, Ahlers HW, Chapin RE, Daston GP, Foster PMD, Kavlock RJ, Morawetz JS, Schnorr TM, Schrader SM [2000]. Prioritization of NTP reproductive toxicants for field studies. Reprod Toxicol 14(4):293-301.

Risk Assessment Methods
Dose Response Models Work Group [2002]. Improving risk assessment: research opportunities in dose response modeling. J Human and Ecological Risk Assessment 8(6):1421.

Epidemiology Work Group [2002]. Improving risk assessment: priorities for epidemiologic research.
J Human and Ecological Risk Assessment 8(6):1397.

Toxicology Work Group [2002]. Improving risk assessment: toxicological research needs. J Human and Ecological Risk Assessment 8(6):1421.

Special Populations at Risk
Frumkin H, Pransky G [1999]. Special populations. Occupational Medicine State of the Art Reviews, 14(3):479-705.

National Research Council [2004]. Health and safety needs of older workers. Washington DC: National Academy Press.

Surveillance Research Methods
Council of State and Territorial Epidemiologists (CSTE) [2001]. The Role of States in a Nationwide Comprehensive Surveillance System for Work-related Diseases, Injuries and Hazards. Atlanta, GA: CSTE.

NIOSH [2001]. Tracking occupational injuries, illnesses, and hazards: the NIOSH Surveillance Strategic Plan. Cincinnati, OH: U.S. Department of Health and Human Services, Centers for Disease Control and Prevention, National Institute for Occupational Safety and Health, DHHS (NIOSH) Publication No. 2001-118.

Traumatic Injuries
NIOSH [1998]. Traumatic occupational injury research needs and priorities: A report by the NORA Traumatic Injury Team. Cincinnati, OH: U.S. Department of Health and Human Services, Centers for Disease Control and Prevention, National Institute for Occupational Safety and Health, DHHS (NIOSH) Publication No. 98-134.

NORA TEAM MEMBERS: 1996-2006

The following is a list of NORA team members and their affiliation(s) during the time of their participation.

Allergic and Irritant Dermatitis	
Stephen P. Berardinelli	CDC/National Institute for Occupational Safety and Health
David I. Bernstein	University of Cincinnati
Raymond Biagini	CDC/National Institute for Occupational Safety and Health
Mark F. Boeniger	CDC/National Institute for Occupational Safety and Health
Carol A. Burnett	CDC/National Institute for Occupational Safety and Health
Richard Fenske	University of Washington
G. Frank Gerberick	The Procter & Gamble Company
James S. Johnson	Lawrence Livermore National Laboratory
Boris D. Lushniak	CDC/National Institute for Occupational Safety and Health
Michael Luster	CDC/National Institute for Occupational Safety and Health
Dino Mattorano	Environmental Protection Agency
Alan N. Moshell	National Institutes of Health
Albert E. Munson	CDC/National Institute for Occupational Safety and Health
Frances Storrs	Oregon Health Sciences University
James S. Taylor	Cleveland Clinic Foundation
Susan Q. Wilburn	American Nurses Association
Asthma and Chronic Obstructive Pulmonary Disease	
John Balmes	University of California San Francisco
William Beckett	University of Rochester
Margaret Becklake	McGill University

Raymond Biagini	CDC/National Institute for Occupational Safety and Health
Robert Castellan	CDC/National Institute for Occupational Safety and Health
Tom Croxton	National Institutes of Health
Paul Enright	University of Arizona
Jeffrey Fedan	CDC/National Institute for Occupational Safety and Health
Paul Henneberger	CDC/National Institute for Occupational Safety and Health
Paul Hewett	CDC/National Institute for Occupational Safety and Health
Suzanne Hurd	National Institutes of Health
Susan Kennedy	University of British Columbia
Hillel Koren	Environmental Protection Agency
Kathleen Kreiss	CDC/National Institute for Occupational Safety and Health
Phillip Morey	AQS Services
Edward Petsonk	CDC/National Institute for Occupational Safety and Health
Steve Redd	Centers for Disease and Prevention
Wayne Sanderson	University of Iowa
Gregory Wagner	CDC/National Institute for Occupational Safety and Health
James Weeks	Advance Technologies and Laboratories International
David Wegman	University of Massachusetts at Lowell
Gail Weinmann	National Institutes of Health
David Weissman	CDC/National Institute for Occupational Safety and Health
Cancer Research Methods	
Steve Bayard	Occupational Safety and Health Administration
Aaron Blair	National Institutes of Health
Paul Brandt-Rauf	Columbia University

Mary Ann Butler	CDC/National Institute for Occupational Safety and Health
David Dankovic	CDC/National Institute for Occupational Safety and Health
Caroline Freeman	Occupational Safety and Health Administration
Irva Hertz-Picciotto	University of California, Davis
Ann F. Hubbs	CDC/National Institute for Occupational Safety and Health
Carol Jones	Mine Safety and Health Administration
Myra Karstadt	Environmental Protection Agency
Gregory L. Kedderis	Chemical Industry Institute of Toxicology
James Melius	New York Laborers' Health & Safety Fund
Ronald Melnick	National Institutes of Health
Carrie A. Redlich	Yale University
Nathaniel Rothman	National Institutes of Health
Avima Ruder	CDC/National Institute for Occupational Safety and Health
Russell E. Savage	CDC/National Institute for Occupational Safety and Health
Mary K. Schubauer-Berigan	CDC/National Institute for Occupational Safety and Health
Paul Schulte	CDC/National Institute for Occupational Safety and Health
Jack Siemiatycki	INRS-Institut Armand-Frappier
Marilyn T. Smith	University of California, Berkeley
Michael Sprinker	United Food and Commercial Workers
Mark Toraason	CDC/National Institute for Occupational Safety and Health
Elizabeth M. Ward	American Cancer Society
Ainsley Weston	CDC/National Institute for Occupational Safety and Health
Control Technology and Personal Protective Equipment	
Stephen P. Berardinelli	CDC/National Institute for Occupational Safety and Health

Donald Campbell	CDC/National Institute for Occupational Safety and Health
David deVries	American Society of Safety Engineers
Richard Duffy	International Association of Fire Fighters
Michael Flynn	University of North Carolina
Bill Heitbrink	University of Iowa
Fred Kissell	CDC/National Institute for Occupational Safety and Health
Bill Kojola	American Federation of Labor- Council of Industrial Organizations
Steve Mallinger	Occupational Safety and Health Administration
Earnest Moyer	CDC/National Institute for Occupational Safety and Health
Larry Reed	CDC/National Institute for Occupational Safety and Health
Mike Seymour	Occupational Safety and Health Administration
Daniel K. Shipp	Industrial Safety Equipment Association
Jenny Topmiller	CDC/National Institute for Occupational Safety and Health
Exposure Assessment Methods	
Martin Abell	CDC/National Institute for Occupational Safety and Health
Tom Armstrong	Exxon Biomedical Sciences, Inc
Carol Boraiko	Middle Tennessee State University
Gayle DeBord	CDC/National Institute for Occupational Safety and Health
Beth Donovan Reh	CDC/National Institute for Occupational Safety and Health
Larry Elliot	CDC/National Institute for Occupational Safety and Health
Dennis Groce	CDC/National Institute for Occupational Safety and Health
Michael J. Keane	CDC/National Institute for Occupational Safety and Health
Paul Knechtges	U.S. Department of Defense
Paul J. Middendorf	CDC/National Institute for Occupational Safety and Health

Dennis O'Brien	INTERNATIONAL UNION, UNITED AUTOMOBILE, AEROSPACE AND AGRICULTURAL IMPLEMENT WORKERS OF AMERICA
David Pegram	U.S. DEPARTMENT OF ENERGY
Christopher Reh	CDC/NATIONAL INSTITUTE FOR OCCUPATIONAL SAFETY AND HEALTH
Bonnie Rogers	UNIVERSITY OF NORTH CAROLINA
Eric Sampson	CENTERS FOR DISEASE AND PREVENTION
Bernard D. Silverstien	CONSULTANT
Gary Spies	MONSANTO COMPANY
Mark Stenzel	EXPOSURE ASSESSMENT APPLICATIONS, LLC
Glenn Talaska	UNIVERSITY OF CINCINNATI
David Utterback	CDC/NATIONAL INSTITUTE FOR OCCUPATIONAL SAFETY AND HEALTH
James Weeks	GEORGE WASHINGTON UNIVERSITY
Ken Williams	CDC/NATIONAL INSTITUTE FOR OCCUPATIONAL SAFETY AND HEALTH
Mary Lynn Woebkenberg	CDC/NATIONAL INSTITUTE FOR OCCUPATIONAL SAFETY AND HEALTH
Emerging Technologies	
Nicholas Ashford	MASSACHUSETTS INSTITUTE OF TECHNOLOGY
James Bartis	RAND CORPORATION
George R. Bockosh	CDC/NATIONAL INSTITUTE FOR OCCUPATIONAL SAFETY AND HEALTH
Kenneth D. Brock	SAFETY MANAGEMENT CONSULTANT
Tai Chan	GENERAL MOTORS CORPORATION
David Conover	CDC/NATIONAL INSTITUTE FOR OCCUPATIONAL SAFETY AND HEALTH
Elaine Cullen	CDC/NATIONAL INSTITUTE FOR OCCUPATIONAL SAFETY AND HEALTH
Maryann D'Alessandro	CDC/NATIONAL INSTITUTE FOR OCCUPATIONAL SAFETY AND HEALTH
Michael Eichberg	SELECT COMMITTEE ON HOMELAND SECURITY
John Finklea	CENTERS FOR DISEASE CONTROL AND PREVENTION AND CENTER TO PROTECT WORKERS RIGHTS

Jack Geissert	WYETH BIOPHARMA
Matt Gillen	CDC/NATIONAL INSTITUTE FOR OCCUPATIONAL SAFETY AND HEALTH
Richard Hartle	CDC/NATIONAL INSTITUTE FOR OCCUPATIONAL SAFETY AND HEALTH
James Jones	CDC/NATIONAL INSTITUTE FOR OCCUPATIONAL SAFETY AND HEALTH
Max Kiefer	CDC/NATIONAL INSTITUTE FOR OCCUPATIONAL SAFETY AND HEALTH
Ray Lovett	WEST VIRGINIA UNIVERSITY
Rafael Moure-Eraso	UNIVERSITY OF MASSACHUSETTS
David Y. Pui	UNIVERSITY OF MINNESOTA
Larry Reed	CDC/NATIONAL INSTITUTE FOR OCCUPATIONAL SAFETY AND HEALTH
Maureen Ruskin	OCCUPATIONAL SAFETY AND HEALTH ADMINISTRATION
Paul Schlecht	CDC/NATIONAL INSTITUTE FOR OCCUPATIONAL SAFETY AND HEALTH
Ted Schoenborn	CDC/NATIONAL INSTITUTE FOR OCCUPATIONAL SAFETY AND HEALTH
Aaron Schopper	CDC/NATIONAL INSTITUTE FOR OCCUPATIONAL SAFETY AND HEALTH
Randal P. Schumacher	SCHUMACHER PARTNERS INTERNATIONAL, LLC
John Sparks	ENVIRONMENTAL PROTECTION AGENCY
Donald J. Stillwell	NATIONAL AERONAUTICS AND SPACE ADMINISTRATION
Jeffrey H. Welsh	CDC/NATIONAL INSTITUTE FOR OCCUPATIONAL SAFETY AND HEALTH
Debra Yu	PFIZER, INC.

Hearing Loss

Rob Brauch	LARSON-DAVIS, INC.
Patricia Brogan	WAYNE STATE UNIVERSITY
David Byrne	CDC/NATIONAL INSTITUTE FOR OCCUPATIONAL SAFETY AND HEALTH
John Casali	VIRGINIA TECH UNIVERSITY
John Franks	CDC/NATIONAL INSTITUTE FOR OCCUPATIONAL SAFETY AND HEALTH

Steven N. Hacker	SOLUTIA INC.
Lee Hager	SONOMAX, INC.
Howard Hoffman	NATIONAL INSTITUTES OF HEALTH
Dan Johnson	ANSI STANDARDS
James E. Lankford	NORTHERN ILLINOIS UNIVERSITY
Anthony Miltich	DIGITAL HEARING SYSTEMS CORPORATION
Doug Ohlin	U.S. ARMY CENTER FOR HEALTH PROMOTION AND PREVENTATIVE MEDICINE
Scott Schneider	CENTER TO PROTECT WORKERS RIGHTS
Carol Merry Stephenson	CDC/NATIONAL INSTITUTE FOR OCCUPATIONAL SAFETY AND HEALTH
Mark R. Stephenson	CDC/NATIONAL INSTITUTE FOR OCCUPATIONAL SAFETY AND HEALTH
Randy Tubbs	CDC/NATIONAL INSTITUTE FOR OCCUPATIONAL SAFETY AND HEALTH

Health Services Research

Joan Buchanan	HARVARD UNIVERSITY
William Bunn	NAVISTAR INTERNATIONAL TRANSPORTATION CORPORATION
Mary A. Cummings	AGENCY FOR HEALTH CARE RESEARCH AND QUALITY
Scott Deitchman	CDC/NATIONAL INSTITUTE FOR OCCUPATIONAL SAFETY AND HEALTH
James Ellenberger	AMERICAN FEDERATION OF LABOR- COUNCIL OF INDUSTRIAL ORGANIZATIONS
Jay Himmelstein	UNIVERSITY OF MASSACHUSETTS
Ted Katz	CDC/NATIONAL INSTITUTE FOR OCCUPATIONAL SAFETY AND HEALTH
Kent W. Peterson	OCCUPATIONAL HEALTH STRATEGIES, INC.
Gordon Reeve	FORD MOTOR COMPANY
Kathleen Rest	CDC/NATIONAL INSTITUTE FOR OCCUPATIONAL SAFETY AND HEALTH
Linda Rudolph	CALIFORNIA DEPARTMENT OF INDUSTRIAL RELATIONS
Gregory Wagner	CDC/NATIONAL INSTITUTE FOR OCCUPATIONAL SAFETY AND HEALTH

Infectious Disease	
James August	AMERICAN FEDERATION OF STATE, COUNTY AND MUNICIPAL EMPLOYEES
Jordan Barab	OCCUPATIONAL SAFETY AND HEALTH ADMINISTRATION
Yvonne A. Boudreau	CDC/NATIONAL INSTITUTE FOR OCCUPATIONAL SAFETY AND HEALTH
Joel Breman	NATIONAL INSTITUTES OF HEALTH
Gwendolyn Cattledge	CDC/NATIONAL INSTITUTE FOR OCCUPATIONAL SAFETY AND HEALTH
Naresh Chawla	U.S. FOOD AND DRUG ADMINISTRATION
George W. Counts	NATIONAL INSTITUTES OF HEALTH
Amy Curtis	CENTERS FOR DISEASE AND PREVENTION
Feroza Daroowalla	CDC/NATIONAL INSTITUTE FOR OCCUPATIONAL SAFETY AND HEALTH
Yvette Davis	CENTERS FOR DISEASE AND PREVENTION
William G. Denton	DENTON INTERNATIONAL
Amanda L. Edens	OCCUPATIONAL SAFETY AND HEALTH ADMINISTRATION
June Fisher	SAN FRANCISCO GENERAL HOSPITAL
Martha Hare	NATIONAL INSTITUTES OF HEALTH
David K. Henderson	NATIONAL INSTITUTES OF HEALTH
Thomas K. Hodous	CDC/NATIONAL INSTITUTE FOR OCCUPATIONAL SAFETY AND HEALTH
Janice Huy	CDC/NATIONAL INSTITUTE FOR OCCUPATIONAL SAFETY AND HEALTH
Paul Jensen	CENTERS FOR DISEASE AND PREVENTION
Kevin Landkrohn	OCCUPATIONAL SAFETY AND HEALTH ADMINISTRATION
June Lunney	NATIONAL INSTITUTES OF HEALTH
Robert Lyerla	CENTERS FOR DISEASE AND PREVENTION
Marissa A. Miller	NATIONAL INSTITUTES OF HEALTH
Robert J. Mullan	CDC/NATIONAL INSTITUTE FOR OCCUPATIONAL SAFETY AND HEALTH

Adelisa L. Panlilio	Centers for Disease and Prevention
Jack Parker	CDC/National Institute for Occupational Safety and Health
Aron Primack	National Institutes of Health
Renee Ridzon	Centers for Disease Control and Prevention and the Bill and Melinda Gates Foundation
Gary Roselle	Veterans Affairs Headquarters
Kevin Seifert	Becton Dickinson and Company
Teresa A. Seitz	CDC/National Institute for Occupational Safety and Health
Kent A. Sepkowitz	Memorial Sloan-Kettering Center
Craig N. Shapiro	Centers for Disease Control and Prevention
Hilary Sigmon	National Institutes of Health
David N. Weissman	CDC/National Institute for Occupational Safety and Health
Ian Williams	Centers for Disease Control and Prevention

Indoor Environments

Daryl Alexander	American Teachers Federation
Denise Bowles	American Federation of State, County and Municipal Employees
Bill Cain	University of California, San Diego
Jean Cox-Ganser	CDC/National Institute for Occupational Safety and Health
William Fisk	Lawrence Berkeley National Laboratory
Caroline Freeman	Occupational Safety and Health Administration
John Girman	Environmental Protection Agency
Cynthia Hines	CDC/National Institute for Occupational Safety and Health
Paul Jenson	CDC/National Institute for Occupational Safety and Health
Kathleen Kreiss	CDC/National Institute for Occupational Safety and Health
Hal Levin	Building Ecology Research Group

Mark Mendell	CDC/National Institute for Occupational Safety and Health and Lawrence Berkeley National Laboratory
Donald Milton	Harvard School of Public Health
Carol Rao	CDC/National Institute for Occupational Safety and Health
Larry Rexroat	U.S. General Services Administration
Mary Smith	Environmental Protection Agency
Eileen Storey	University of Connecticut
Ken Wallingford	CDC/National Institute for Occupational Safety and Health
Martha Waters	CDC/National Institute for Occupational Safety and Health
Ray Wells	CDC/National Institute for Occupational Safety and Health
Charles Weschler	University of Medicine and Dentistry of New Jersey and the Robert Wood Johnson Medical School
Intervention Effectiveness Research	
Robin Baker	University of California, Berkeley
Denis Bourcier	The Boeing Company
Ann Brockhaus	Organization Resource Counselors, Inc.
Larry Chapman	University of Wisconsin
Patrick Coleman	CDC/National Institute for Occupational Safety and Health
Jim Collins	CDC/National Institute for Occupational Safety and Health
Scott Earnest	CDC/National Institute for Occupational Safety and Health
Sarah Glavin	The Government Accountability Office
Linda Goldenhar	CDC/National Institute for Occupational Safety and Health
Amanda Harney	CDC/National Institute for Occupational Safety and Health
Cathy Heaney	Ohio State University
Janet Johnston	CDC/National Institute for Occupational Safety and Health
Ted Katz	CDC/National Institute for Occupational Safety and Health

Tony LaMontagne	NEW ENGLAND RESEARCH INSTITUTES
Paul Landsbergis	CORNELL UNIVERSITY AND MT. SINAI MEDICAL COLLEGE
John Martonik	OCCUPATIONAL SAFETY AND HEALTH ADMINISTRATION
Ivan Most	
Daniel Murphy	ST. PAUL FIRE AND MARINE INSURANCE COMPANY
Carolyn Needleman	BRYN MAWR COLLEGE
Ted Scharf	CDC/NATIONAL INSTITUTE FOR OCCUPATIONAL SAFETY AND HEALTH
Scott Schneider	CENTER TO PROTECT WORKERS RIGHTS AND LABORER'S UNION
Paul Schulte	CDC/NATIONAL INSTITUTE FOR OCCUPATIONAL SAFETY AND HEALTH
Ray Sinclair	CDC/NATIONAL INSTITUTE FOR OCCUPATIONAL SAFETY AND HEALTH
Naomi G. Swanson	CDC/NATIONAL INSTITUTE FOR OCCUPATIONAL SAFETY AND HEALTH
Margaret Wallace	CDC/NATIONAL INSTITUTE FOR OCCUPATIONAL SAFETY AND HEALTH
Mixed Exposures	
Nancy Bollinger	CDC/NATIONAL INSTITUTE FOR OCCUPATIONAL SAFETY AND HEALTH
John Bucher	NATIONAL INSTITUTES OF HEALTH
Gregory Burr	CDC/NATIONAL INSTITUTE FOR OCCUPATIONAL SAFETY AND HEALTH
Vincent Castranova	CDC/NATIONAL INSTITUTE FOR OCCUPATIONAL SAFETY AND HEALTH
Gregory L. Finch	PFIZER, INC.
Hank Gardner	COLORADO STATE UNIVERSITY
Manuel Gomez	AMERICAN INDUSTRIAL HYGIENE ASSOCIATION
Judith Graham	AMERICAN CHEMISTRY COUNCIL
Hugh Hansen	CENTERS FOR DISEASE CONTROL AND PREVENTION
Frank Hearl	CDC/NATIONAL INSTITUTE FOR OCCUPATIONAL SAFETY AND HEALTH
Robert Herrick	HARVARD UNIVERSITY

Richard Hertzberg	U.S. ENVIRONMENTAL PROTECTION AGENCY
Mark D. Hoover	CDC/NATIONAL INSTITUTE FOR OCCUPATIONAL SAFETY AND HEALTH
Vera Kommineni	CDC/NATIONAL INSTITUTE FOR OCCUPATIONAL SAFETY AND HEALTH
Daniel Lewis	CDC/NATIONAL INSTITUTE FOR OCCUPATIONAL SAFETY AND HEALTH
Alan Lunsford	CDC/NATIONAL INSTITUTE FOR OCCUPATIONAL SAFETY AND HEALTH
Margaret MacDonell	DEPARTMENT OF ENERGY
Daniel J. Marsick	U.S. DEPARTMENT OF ENERGY
Joe L. Mauderly	LOVELACE RESPIRATORY RESEARCH INSTITUTE
Moiz Mumtaz	CENTERS FOR DISEASE CONTROL AND PREVENTION
Guenter Oberdoerster	UNIVERSITY OF ROCHESTER
William G. Perry	OCCUPATIONAL SAFETY AND HEALTH ADMINISTRATION
Peter Robinson	MANTECH INC. AND U.S. DEPARTMENT OF DEFENSE
Sharon Silver	CDC/NATIONAL INSTITUTE FOR OCCUPATIONAL SAFETY AND HEALTH
Pam Susi	CENTER TO PROTECT WORKERS RIGHTS
Douglas Trout	CDC/NATIONAL INSTITUTE FOR OCCUPATIONAL SAFETY AND HEALTH
Raymond Yang	COLORADO STATE UNIVERSITY
Musculoskeletal Disorders	
Stephen Brightwell	CDC/NATIONAL INSTITUTE FOR OCCUPATIONAL SAFETY AND HEALTH
Susan Burt	CDC/NATIONAL INSTITUTE FOR OCCUPATIONAL SAFETY AND HEALTH
LaMont Byrd	INTERNATIONAL BROTHERHOOD OF TEAMSTERS
Greg Cutlip	CDC/NATIONAL INSTITUTE FOR OCCUPATIONAL SAFETY AND HEALTH
Cheryl Fairfield Estill	CDC/NATIONAL INSTITUTE FOR OCCUPATIONAL SAFETY AND HEALTH
Lawrence Fine	NATIONAL INSTITUTES OF HEALTH
Tom Leamon	LIBERTY MUTUAL RESEARCH CENTER FOR SAFETY AND HEALTH

William Marras	Ohio State University
David May	Occupational Safety and Health Administration
Lida Orta-Anes	Graduate School of Public Health University of Puerto Rico
Jim Panagis	National Institutes of Health
Kellie Pierson	CDC/National Institute for Occupational Safety and Health
Vernon Putz Anderson	CDC/National Institute for Occupational Safety and Health
Barbara Silverstein	SHARP, Department of Labor & Industries
Tom Slavin	International Truck and Engine Corporation
Thomas R. Waters	CDC/National Institute for Occupational Safety and Health

Organization of Work

Eileen Appelbaum	Rutgers University
W. Stephen Brightwell	CDC/National Institute for Occupational Safety and Health
Tim Bushnell	CDC/National Institute for Occupational Safety and Health
Claire C. Caruso	CDC/National Institute for Occupational Safety and Health
Michael J. Colligan	CDC/National Institute for Occupational Safety and Health
Donald Eggerth	CDC/National Institute for Occupational Safety and Health
Maryella Gockel	Ernst & Young, LLP
Amanda Harney	CDC/National Institute for Occupational Safety and Health
Anneke Heitmann	Circadian Technologies, Inc
Joseph J. Hurrell, Jr.	CDC/National Institute for Occupational Safety and Health
Theodore M. Katz	CDC/National Institute for Occupational Safety and Health
Bill Kojola	American Federation of Labor- Council of Industrial Organizations
David LeGrande	Communications Workers of America
Nancy Lessin	Massachusetts American Federation of Labor- Council of Industrial Organizations

Richard A. Lippin	USA MEDDAC
Jane A. Lipscomb	UNIVERSITY OF MARYLAND
Vicki Magley	UNIVERSITY OF CINCINNATI
Lawrence R. Murphy	CDC/NATIONAL INSTITUTE FOR OCCUPATIONAL SAFETY AND HEALTH
Katharine Newman	BUREAU OF LABOR STATISTICS
Jeannie Nigam	CDC/NATIONAL INSTITUTE FOR OCCUPATIONAL SAFETY AND HEALTH
Robert H. Peters	CDC/NATIONAL INSTITUTE FOR OCCUPATIONAL SAFETY AND HEALTH
Gwendolyn Puryear Keita	AMERICAN PSYCHOLOGICAL ASSOCIATION
Sydney R. Robertson	ORGANIZATIONAL RESOURCES COUNSELORS, INC
Roger Rosa	CDC/NATIONAL INSTITUTE FOR OCCUPATIONAL SAFETY AND HEALTH
Steven L. Sauter	CDC/NATIONAL INSTITUTE FOR OCCUPATIONAL SAFETY AND HEALTH
Anita Schill	CDC/NATIONAL INSTITUTE FOR OCCUPATIONAL SAFETY AND HEALTH
Jeanne Mager Stellman	COLUMBIA UNIVERSITY
Naomi Swanson	CDC/NATIONAL INSTITUTE FOR OCCUPATIONAL SAFETY AND HEALTH
Lois E. Tetrick	UNIVERSITY OF HOUSTON
Bryan Vila	U.S. DEPARTMENT OF JUSTICE
Deborah Weinstock	AMERICAN FEDERATION OF LABOR- COUNCIL OF INDUSTRIAL ORGANIZATIONS

Risk Assessment Methods

Michael Attfield	CDC/NATIONAL INSTITUTE FOR OCCUPATIONAL SAFETY AND HEALTH
David Dankovic	CDC/NATIONAL INSTITUTE FOR OCCUPATIONAL SAFETY AND HEALTH
Elaine Faustman	UNIVERSITY OF WASHINGTON
William Fayerweather	OWENS CORNING CORPORATION
Herman Gibb	ENVIRONMENTAL PROTECTION AGENCY
Dale Hattis	CLARK UNIVERSITY

Annie Jarabek	ENVIRONMENTAL PROTECTION AGENCY
Frank Mirer	INTERNATIONAL UNION, UNITED AUTOMOBILE, AEROSPACE AND AGRICULTURAL IMPLEMENT WORKERS OF AMERICA
William G. Perry	OCCUPATIONAL SAFETY AND HEALTH ADMINISTRATION
Christopher J. Portier	NATIONAL INSTITUTES OF HEALTH
Val (Skip) Schaeffer	OCCUPATIONAL SAFETY AND HEALTH ADMINISTRATION
Christine Sofge	CDC/NATIONAL INSTITUTE FOR OCCUPATIONAL SAFETY AND HEALTH
Leslie Stayner	UNIVERSITY OF ILLINOIS
Kyle Steenland	EMORY UNIVERSITY
Jane Teta	EXPONENT CORPORATION
Mark Toraason	CDC/NATIONAL INSTITUTE FOR OCCUPATIONAL SAFETY AND HEALTH
Gerald Van Belle	UNIVERSITY OF WASHINGTON
Terry Wassell	CDC/NATIONAL INSTITUTE FOR OCCUPATIONAL SAFETY AND HEALTH
Ainsley Weston	CDC/NATIONAL INSTITUTE FOR OCCUPATIONAL SAFETY AND HEALTH

Reproductive Health Research (Fertility & Pregnancy Abnormalities)

Colleen Boyle	CENTERS FOR DISEASE CONTROL AND PREVENTION
Ken Bridbord	NATIONAL INSTITUTES OF HEALTH
John Cook	PFIZER, INC.
Sally Perreault Darney	ENVIRONMENTAL PROTECTION AGENCY
George Daston	PROCTOR AND GAMBLE
Linda Frazier	UNIVERSITY OF KANSAS
Barbara Grajewski	CDC/NATIONAL INSTITUTE FOR OCCUPATIONAL SAFETY AND HEALTH
James Kesner	CDC/NATIONAL INSTITUTE FOR OCCUPATIONAL SAFETY AND HEALTH
Christina Lawson	CDC/NATIONAL INSTITUTE FOR OCCUPATIONAL SAFETY AND HEALTH
Dennis Lynch	CDC/NATIONAL INSTITUTE FOR OCCUPATIONAL SAFETY AND HEALTH

Barbara MacKenzie	CDC/NATIONAL INSTITUTE FOR OCCUPATIONAL SAFETY AND HEALTH
Michele Marcus	EMORY UNIVERSITY
Donald R. Mattison	NATIONAL INSTITUTES OF HEALTH
Melissa McDiarmid	UNIVERSITY OF MARYLAND
Eisuke Murono	CDC/NATIONAL INSTITUTE FOR OCCUPATIONAL SAFETY AND HEALTH
Wendie A. Robbins	UNIVERSITY OF CALIFORNIA, LOS ANGELES
Margaret A.K. Ryan	U.S. NAVY
Teresa Schnorr	CDC/NATIONAL INSTITUTE FOR OCCUPATIONAL SAFETY AND HEALTH
Steven Schrader	CDC/NATIONAL INSTITUTE FOR OCCUPATIONAL SAFETY AND HEALTH
Michael Shelby	NATIONAL INSTITUTES OF HEALTH
Elizabeth A. Whelan	CDC/NATIONAL INSTITUTE FOR OCCUPATIONAL SAFETY AND HEALTH

Social and Economic Consequences of Workplace Injury and Illness

Benjamin C. Amick III	UNIVERSITY OF TEXAS
Elyce Biddle	CDC/NATIONAL INSTITUTE FOR OCCUPATIONAL SAFETY AND HEALTH
Laura Blanciforti	CDC/NATIONAL INSTITUTE FOR OCCUPATIONAL SAFETY AND HEALTH
Leslie I. Boden	BOSTON UNIVERSITY
Thomas W. Camm	CDC/NATIONAL INSTITUTE FOR OCCUPATIONAL SAFETY AND HEALTH
Brian T. Day	CDC/NATIONAL INSTITUTE FOR OCCUPATIONAL SAFETY AND HEALTH
Ross Eisenbrey	OCCUPATIONAL SAFETY AND HEALTH ADMINISTRATION
Theodore M. Katz	CDC/NATIONAL INSTITUTE FOR OCCUPATIONAL SAFETY AND HEALTH
Paul R. Keane	CDC/NATIONAL INSTITUTE FOR OCCUPATIONAL SAFETY AND HEALTH
Chuck Levenstein	UNIVERSITY OF MASSACHUSETTS, LOWELL
Stephen Luchter	NATIONAL HIGHWAY TRAFFIC SAFETY ADMINISTRATION
Ted Miller	PACIFIC INSTITUTES FOR RESEARCH AND EVALUATION

Sharon Morris	UNIVERSITY OF WASHINGTON
Robert Reville	THE RAND CORPORATION
Harry Shannon	MCMASTER UNIVERSITY
Eve Spangler	BOSTON COLLEGE
Emily Spieler	WEST VIRGINIA UNIVERSITY
Special Populations at Risk	
Sherry Baron	CDC/NATIONAL INSTITUTE FOR OCCUPATIONAL SAFETY AND HEALTH
Joni Berardino	NATIONAL CENTER FOR FARMWORKER HEALTH
Cecil Buzz Burchfiel	CDC/NATIONAL INSTITUTE FOR OCCUPATIONAL SAFETY AND HEALTH
Lorri Cameron	CDC/NATIONAL INSTITUTE FOR OCCUPATIONAL SAFETY AND HEALTH
Dawn Castillo	CDC/NATIONAL INSTITUTE FOR OCCUPATIONAL SAFETY AND HEALTH
Gwendolyn Cattledge	CDC/NATIONAL INSTITUTE FOR OCCUPATIONAL SAFETY AND HEALTH
Letitia Davis	MASSACHUSETTS DEPARTMENT OF PUBLIC HEALTH
Gayle DeBord	CDC/NATIONAL INSTITUTE FOR OCCUPATIONAL SAFETY AND HEALTH
Anne Fidler	CDC/NATIONAL INSTITUTE FOR OCCUPATIONAL SAFETY AND HEALTH
Howard Frumkin	NATIONAL INSTITUTES OF HEALTH
Jim Grosch	CDC/NATIONAL INSTITUTE FOR OCCUPATIONAL SAFETY AND HEALTH
Paul H. Grundy	IBM
Frances C. Henderson	ALCORN STATE UNIVERSITY
James Jackson	UNIVERSITY OF MICHIGAN
Charles Lee	UNITED CHURCH OF CHRIST
Leslie MacDonald	CDC/NATIONAL INSTITUTE FOR OCCUPATIONAL SAFETY AND HEALTH
Theodore Meinhardt	CDC/NATIONAL INSTITUTE FOR OCCUPATIONAL SAFETY AND HEALTH
Jacqueline Nowell	UNITED FOOD AND COMMERCIAL WORKERS

Scott Richardson	BUREAU OF LABOR STATISTICS
Richard D. Rinehart	OCCUPATIONAL SAFETY AND HEALTH ADMINISTRATION
Rosemary Sokas	CDC/NATIONAL INSTITUTE FOR OCCUPATIONAL SAFETY AND HEALTH AND UNIVERSITY OF ILLINOIS CHICAGO
Sandra Stratford	IBM
Pamela Tau Lee	LABOR OCCUPATIONAL HEALTH PROGRAM
Andrea Taylor	INTERNATIONAL UNION, UNITED AUTOMOBILE, AEROSPACE AND AGRICULTURAL IMPLEMENT WORKERS OF AMERICA
Luz Maritza Tennassee	PAN AMERICAN HEALTH ASSOCIATION
Rex Tingle	OCCUPATIONAL SAFETY AND HEALTH ADMINISTRATION
Jennie Ward Robinson	CDC/NATIONAL INSTITUTE FOR OCCUPATIONAL SAFETY AND HEALTH
David Wegman	UNIVERSITY OF MASSACHUSETTS, LOWELL
Anthony Yancey	CDC/NATIONAL INSTITUTE FOR OCCUPATIONAL SAFETY AND HEALTH

Surveillance Research Methods

Robert Castellan	CDC/NATIONAL INSTITUTE FOR OCCUPATIONAL SAFETY AND HEALTH
Joe Dubois	OCCUPATIONAL SAFETY AND HEALTH ADMINISTRATION
Janet Ehlers	CDC/NATIONAL INSTITUTE FOR OCCUPATIONAL SAFETY AND HEALTH
Richard Ehrenberg	CDC/NATIONAL INSTITUTE FOR OCCUPATIONAL SAFETY AND HEALTH
Barbara Fotta	CDC/NATIONAL INSTITUTE FOR OCCUPATIONAL SAFETY AND HEALTH
Eric Frumin	UNITE
Manuel Gomez	AMERICAN INDUSTRIAL HYGIENE ASSOCIATION
Alice Greife	CENTRAL MISSOURI STATE UNIVERSITY
Wayne Lednar	EASTMAN KODAK
Jane McCammon	CDC/NATIONAL INSTITUTE FOR OCCUPATIONAL SAFETY AND HEALTH
Leroy Mickelsen	CDC/NATIONAL INSTITUTE FOR OCCUPATIONAL SAFETY AND HEALTH
Tim Morse	UNIVERSITY OF CONNECTICUT HEALTH CENTER

John Myers	CDC/National Institute for Occupational Safety and Health
Steve Newell	Organizational Resources Counselors, Inc.
Kara Perritt	CDC/National Institute for Occupational Safety and Health
Earl Pollack	Center to Protect Workers Rights
Lee Sanderson	CDC/National Institute for Occupational Safety and Health
Anita Schill	CDC/National Institute for Occupational Safety and Health
Patricia Schleiff	CDC/National Institute for Occupational Safety and Health
John Sestito	CDC/National Institute for Occupational Safety and Health
Traumatic Injuries	
Lani Boldt	CDC/National Institute for Occupational Safety and Health
Bill Borwegen	Service Employers International Union
Christine Branche	Centers for Disease Control and Prevention
George Conway	CDC/National Institute for Occupational Safety and Health
Bob DeSiervo	American Society of Safety Engineers
Timothy Fisher	American Society of Safety Engineers
Jim Harris	CDC/National Institute for Occupational Safety and Health
Alan Hoskin	National Safety Council
Lynn Jenkins	CDC/National Institute for Occupational Safety and Health
Herb Linn	CDC/National Institute for Occupational Safety and Health
Stephen Luchter	National Highway Traffic Safety Administration
Jan Manwaring	CDC/National Institute for Occupational Safety and Health
Nancy McWilliams	McWilliams Risk Management
Jacqueline Nowell	United Food and Commercial Workers
Corinne Peek-Asa	University of Iowa

Tim Pizatella	CDC/National Institute for Occupational Safety and Health
Gordon Reeve	Ford Motor Company
Gordon Smith	Johns Hopkins University and Liberty Mutual Center for Injury Research
Karl Snyder	CDC/National Institute for Occupational Safety and Health
Gary Sorock	Liberty Mutual Research Center
Lisa Steiner	CDC/National Institute for Occupational Safety and Health
Nancy Stout	CDC/National Institute for Occupational Safety and Health

www.ingramcontent.com/pod-product-compliance
Lightning Source LLC
Chambersburg PA
CBHW051805170526
45167CB00005B/1891